Edited by Jon Elster and Aanund Hylland

FOUNDATIONS OF SOCIAL CHOICE THEORY

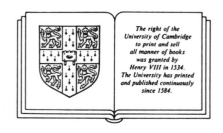

The right of the
University of Cambridge
to print and sell
all manner of books
was granted by
Henry VIII in 1534.
The University has printed
and published continuously
since 1584.

Cambridge University Press
Cambridge
New York – Port Chester – Melbourne – Sydney

Universitetsforlaget
Oslo – Bergen – Stavanger – Tromsø

Published in collaboration with Maison des Sciences de l'Homme, Paris

Published by the Press Syndicate of the University of Cambridge
The Pitt Building, Trumpington Street, Cambridge CB2 1RP
40 West 20th Street, New York, NY 10011, USA
10 Stamford Road, Oakleigh, Melbourne 3166, Australia

© Cambridge University Press and Universitetsforlaget 1986

First published 1986
Reprinted 1987
First paperback edition 1989

Printed in the United States of America

Library of Congress catalogue card number: 84-29248

British Library Cataloguing in Publication Data
Foundations of social choice theory. – (Studies
in rationality and social change)
1. Social choice
I. Elster, Jon   II. Hylland, Aanund   III. Maison
des sciences de l'homme   IV. Series
302.1'3   HB846.8

ISBN 0-521-25735-2 hard covers
ISBN 0-521-38913-5 paperback

Contents

Preface

The origin of this volume was a conference on 'The foundations of social choice theory' that took place at Ustaoset, Norway in January 1981. There is no direct correspondence, however, between the papers read at the conference and the present volume—neither with respect to the contributors nor with respect to their contributions. We regret that Amos Tversky did not have the occasion to revise for publication his contribution to the conference. As a poor second-best, some of the central ideas in his paper are briefly set out in Elster's contribution.

The conference was sponsored by the Norwegian Research Council for Science and the Humanities, as part of the research project 'Democracy and Social Planning', and by the Maison des Sciences de l'Homme (Paris) as part of an ongoing 'Working Group on Rationality'. We are very grateful to these institutions for their generous assistance.

J.E. and Aa.H.
Oslo, August 1983

Notes on contributors

Brian Barry is Professor of Philosophy at the California Institute of Technology

Donald Davidson is Professor of Philosophy at the University of California, Berkeley

Jon Elster is Professor of Political Science at the University of Chicago and Research Director, Institute for Social Research, Oslo

Allan Gibbard is Professor of Philosophy at the University of Michigan

Robert Goodin is Senior Lecturer in Government at the University of Essex

Aanund Hylland is Senior Lecturer in Economics at the University of Oslo

John Roemer is Professor of Economics at the University of California, Davis

Amartya Sen is Drummond Professor of Political Economy at the University of Oxford

Introduction

JON ELSTER AND AANUND HYLLAND

Any body of scientific knowledge may grow in two directions. On the one hand it may expand upwards and outwards, adding ever more branches to a common trunk of assumptions and axioms. On the other hand it may endeavour to strike its roots ever more deeply, or, to change the metaphor, to secure more solid foundations for the superstructure. In every discipline, these seem to be alternate stages of development. After a period of expansion, there follows a period of consolidation and rethinking that in turn enables expansion along new lines. The inward-looking turn can be provoked in several ways. The theory may encounter anomalies, counter-intuitive or counter-empirical results. It may suffer decreasing marginal productivity. Or it may increasingly lose touch with the problems that originally constituted its *raison d'être*. This last danger is especially acute in the formal sciences, like mathematics. As John von Neumann once remarked, mathematics out of touch with physical sciences tends to become *baroque*, a term used in contrast to the *classical* style of thinking that is constantly revitalized by contact with the empirical sciences.

Formal theorizing in the social sciences is today in some danger of becoming baroque. A frequent scenario seems to be the following. In a first stage, there exists a theoretical problem with immediate economic, social or political significance. It is, however, ill-understood, perhaps even ill-defined. In the second stage, a proposal is put forward to conceptualize the problem in a way that dispels confusion and permits substantive conclusions to be drawn. In a third stage the conceptual apparatus ceases to have these liberating effects, and becomes a new, independent source of problems. An illustration could be the notion of supply and demand schedules. Before the invention of this simple diagrammatic device, the notion of a *change* in supply or demand was hopelessly confused. With this tool, on the other hand, the distinction between price-induced changes and changes induced by a shift in the schedules followed immediately. In a further stage, the refinement of the concepts led to the propositions of general equilibrium theory, which has at best a very tenuous connection with actual market operations.

Social choice theory sprang from two distinct problems: that of finding an adequate, robust voting system, and that of finding a measure for aggregate social welfare. The very idea that there was a close relation between these problems was not clearly perceived. As usual, a notational breakthrough was decisive. By stating the problem as one of finding a *function from a set of individual preferences to a social preference order*, connections could be seen and precise questions could be asked that earlier were only an inchoate possibility. Following Kenneth Arrow's pioneering work, major results were proved at a gratifying rate. Today, social choice theory may be approaching the baroque stage. Breakthroughs are dwindling, while minor embellishments are accelerating. Formalism is gaining the upper hand, as in what Ragnar Frisch used to refer to as 'playometrics'. From a means, formal modelling is becoming an end in itself. These comments are not intended as dismissive of the extensions and refinements that are currently being produced. Indeed, in many cases it is only by stretching the concepts to the limits that their weak points appear. The interplay between intensive and extensive development is part and parcel of scientific development, and it would be absurd to assign one to a higher position than the other. The time may nevertheless be appropriate for a new look at the foundations of social choice theory. The present volume, to be sure, can only survey a few of the issues, and, moreover, very inadequately. It is hoped that it will contribute as much to shifting the emphasis of discussion towards foundational issues as to illuminating the specific problems raised in the various contributions.

When discussing foundations, one may proceed in several ways. First, one may examine the relation between a given discipline and adjoining fields, in the hope that generalizations and simplifications will emerge. Secondly, one may concentrate on the substantive interpretation of the formal results, with a view to finding inspiration for new theoretical developments. Thirdly, one may reconsider the basic assumptions—implicit and explicit—of the theory, in order to find and strengthen the weak links in the chain. The contributions to the present volume illustrate the second and third procedures, while the first is largely neglected. In what follows, the first and second are covered rather briefly, while the third is the subject of somewhat more extensive comment. In particular, we discuss some possible *self-supporting features* of aggregation mechanisms.

Lateral connections

Social choice theory has obvious links to cooperative and non-cooperative game theory. Also, the connection with theories of distributive justice is increasingly intimate. The structure of the theory has been considerably clarified by the explicit consideration of these connections.

Social choice theory and cooperative game theory have many common features. In particular, *n*-person bargaining theory and social choice theory are both concerned with deriving a social outcome from individual preferences, with Pareto optimality as one main constraint. The arbitrator (in bargaining theory) or the constitutional designer (in social choice theory) ask themselves what Pareto-optimal outcome ought to be preferred, given what they know about individual preferences and their notions of how the outcome ought to reflect or respect them. These notions have, in both theories, an essential counterfactual element. They do not simply postulate relations between the actual preferences of the actual individuals over the actual alternatives, on the one hand, and the socially preferred outcome on the other. They also impose consistency conditions on how the preferred outcome ought to change when we vary the preferences of the individuals, the number of individuals, or the set of feasible alternatives. Such consistency is a necessary if not sufficient condition of both rationality and morality.

These broadly similar features are overshadowed, however, by the many dissimilarities in the framework of the theories. Bargaining theories impose more structure on the problem than the standard Arrowian social choice theory, with respect both to the feasible set and to the individual preferences. Also, the outcome to a bargaining problem is supposed to reflect the bargaining strength of the participants, i.e. what would happen to them if they failed to reach agreement. There is no analogous feature in the formulation of the social choice problem.[1]

Non-cooperative game theory seemed rather foreign to Arrow's original statement, but has recently become increasingly central. This shift in emphasis.went together with increasing attention to the fact that expressing inputs to the social decision mechanism is an *action*, one which is guided by the preferences of the individual. He may or may not find it in his interests, as defined by his 'real preferences', to use the latter as his input to the decision process. Clearly, the question of what preferences to express is a problem for non-cooperative game theory, since the answer may depend on what other people can be expected to do. Specifically, one may ask under what circumstances, if any, honest voting can be a dominant strategy; more weakly, whether the voting game has equilibria in which all players express their real preferences; moreover, whether the solution to the game is an equilibrium point of this type.[2] Similarly, whether the agents should use or waive their *rights* turns out to be an important question in some formulations of the Liberal Dilemma. This is also a game-theoretic issue, since the decision may depend on whether others can be expected to waive *their* rights.

Arrow's theory has been characterized by himself as 'ordinal utilitarianism', i.e. a conception of distributive justice. It may then be usefully contrasted with other conceptions of this kind, e.g. classical utilitarianism and Rawls's difference principle. One of the more important developments in social choice theory has been the proof that by an extension of its framework both the latter theories can be subsumed as special cases. Specifically, by allowing more utility information than is possible within the ordinal framework, one can give axiom sets that uniquely characterize the average utility or the minimum utility as the proper maximand for the social welfare function.[3] This broad group of conceptions may be contrasted with those which use bargaining theory as the proper framework for settling issues of distributive justice.[4]

This cluster of theoretical clusters—games, bargaining, justice, and

---

[1] For discussions of the relation between bargaining theory and social choice theory, see Luce and Raiffa (1957, pp. 349 ff.) and Sen (1970, Ch. 8).

[2] See Dasgupta, Hammond, and Maskin (1979) for an overview.

[3] See d'Aspremont and Gevers (1977).

[4] See, for instance, Gauthier (forthcoming).

social choice—exhibits all sorts of intricate internal connections. Problems from one cluster can often be reformulated in the language of another. Yet, we are a far cry from the transparency and simplicity that would characterize a truly unified theory. Indeed, unification may be a will-o'-the-wisp, although one may at least expect partial simplifications and generalizations to occur as the result of further cross-disciplinary work.

Interpretations

Any formal structure can be interpreted in several distinct ways. One specific interpretation is often that intended by the creators of the structure. Later, it is seen that other, substantively different problems can be subsumed under the same set of axioms. These problems may suggest additions to or modifications of the formal structure, and thus offer new insight into the original set of problems. This interplay between substantive issues and formal modelling is important in securing better foundations for the theory.

In the social sciences and in philosophy there is one particular instance of this interplay that has been especially important. When studying some kind of interaction *between* persons, from a normative or a behavioural point of view, it is often instructive to look into whether a similar interaction can be observed *within* a single person. It will often be found that the concepts developed for the interpersonal case also apply to the intrapersonal one, but not in a wholly unstrained way. To improve the fit, changes are needed that, on further reflection, may also throw light on the original interpersonal problem.

In social choice theory this extension to intrapersonal cases can occur in several ways. One is instantiated in the contribution of Ian Steedman and Ulrich Krause to *The Multiple Self*, a companion volume to the present work. They observe that an individual often evaluates a choice situation from many different perspectives, each of which has associated with it a preference ordering of the options. My egoistic, altruistic, moral and social selves may rank the alternatives differently, so that the need arises for some aggregation mechanism. Some features of standard social choice theory appear somewhat differently in this setting. For instance, there might not be anything objectionable in having a dictator. We might want to say that the moral preference structure ought to win out in all cases. Yet, since we know that this outcome is far from always observed in reality, we may reformulate the question as a second-best problem: given that some intrapersonal democracy is unavoidable, how ought it to be set up? Another intrapersonal extension is to intertemporal decisions. We may conceive of each time slice of the individual as a separate 'self' with interests that range forward into the future and backward into the past. This procedure would enable us to ask questions such as the following: Ought there to be a condition of 'liberalism' that would block preferences among past options to influence the choice between them? That is, given two consumption streams that are identical from time t onwards, should we allow post-t selves any say in defining the preference between them?

In addition to the distinction between interpersonal and intrapersonal applications, there is the better-known distinction between interprofile and intraprofile applications. The former involve consistency conditions of a counterfactual sort, whereas the latter consider just one actual preference profile. Impossibility theorems may be proved for both cases,[5] although of course one would have to be careful when interpreting the terms. For instance, the existence of a 'dictator' has a very special meaning in the intraprofile case, since it cannot have any implication about *power*, which is an essentially counterfactual notion. By contrast, in the interprofile case a dictator is defined as someone who gets his way whatever his preferences and those of all others might happen to be. By exploring partial analogies of these kinds, we acquire a firmer grasp of what was involved in the original theory, and may even be motivated to modify it to incorporate some of the features from the analogous cases.

Assumptions

In social choice theory under the standard (interpersonal and interprofile) interpretation, various assumptions are made explicitly or implicitly. Here we focus on some of the unstated assumptions, to see whether an explicit consideration can suggest some further lines of research. Most importantly, the framework takes a number of features as exogenously given that are in reality the outcome of complex

[5] Roberts (1980).

economic, social and political processes. Also, whether the social decision mechanism ought not itself to be socially chosen is a question frequently asked, but rarely answered.

Among the features usually taken as given are (i) the set of individuals, (ii) their preferences, (iii) the options confronting them, and (iv) the rights assigned to them. In the present volume, the contributions by Elster and Goodin consider in some detail the question of preference formation and preference filtering. The question of endogenous agenda formation is receiving a good deal of attention in the literature on voting systems. This is closely related to the problems of strategic expression of preferences. (Given the Borda rule, for example, one may distort the decision process either by misrepresenting one's preferences or by adding items to the agenda.) In what follows we consider the endogenous formation of the political unit, the endogenous formation of rights, and the endogenous emergence of the social choice mechanism itself. The discussion of these various questions is not, of course, presuppositionless. It rests on further assumptions that may well be questioned. Also, it is formulated within a less general framework than the standard Arrowian theory. Since the intention is mainly to illustrate a line of argument, this does not matter too much.

Much of politics concerns the distribution of burdens and benefits within a given political unit. Throughout history, however, the question of where to draw the boundaries between political units has also been of overwhelming importance. For social choice theory the problem is whether it is possible to suggest a normative justification for drawing these boundaries in one way rather than in another. We approach this question somewhat indirectly. We first mention a couple of respects in which a political system might be said to be self-supporting or self-justifying, and then go on to ask whether a similar argument could hold for the drawing of boundaries.

Consider first the question of suffrage. In any polity, some individuals lack the right to vote: children, criminals, foreign citizens, etc. The question then arises as to how the right to vote is to be assigned. There may not exist a general answer, but in one particular case it seems possible to use consistency arguments to reach a solution. This is the case of voting age. Let us assume that we ask all individuals more than x years old what, in their opinion, ought to be the voting age.

Specifically, for each n we ask them whether persons at least n years old ought to have the right to vote. The smallest n such that there is a simple majority to let people at least n years old have the right to vote, we call $f(x)$. It is highly likely that, say, $f(6)>6$ and $f(50)<50$. Also, it is probable that for all x, $f(x + 1) \geqslant f(x)$. This ensures that there is some x such that $f(x)$ approximately equals x. We submit that the voting age ought to be the smallest x with this property. The proposal assumes that democracy ought to be extended as far as can consistently be defended. Thus, if in the group above 16 years there is a majority against letting 16-year-olds have the right to vote, this ought not to be chosen as the voting age. Conversely, one ought not to set the voting age at 21 if there is a majority among persons above 18 years of age for the view that people above 18 ought to have the right to vote.

Consider next the question of majority voting. For various reasons one may sometimes desire to use qualified majority voting on fundamental constitutional issues. The obvious question is: how does one decide which majority to require? In the light of the preceding paragraph, the answer should be clear. At any given time there is a well-defined percentage $g(x)$ of the voters that want the majority required for a certain constitutional change to be at least x per cent. $g(x)$ must be a decreasing function of x; also we may assume $g(50)>50$ and $g(100)<100$. Then, again, we ought to choose a self-supporting majority: the unique $\bar{x}$ such that $g(\bar{x}) = \bar{x}$.

Against the background of these two examples, we may consider the question of boundary-drawing. The obvious suggestion now is the following. One ought to define as one political unit the largest territorial unit such that within the unit there is a simple majority for considering it as one voting community rather than several. It may also be obvious why this proposal doesn't work. Unlike the other examples, one cannot make the crucial monotonicity assumption that allows the argument to go through. One cannot assume, that is, that the feeling of belonging to one community becomes progressively weaker as one moves outward from the centre. The following example suffices to prove the point:

Consider, for example, the Irish question as it stood between 1918 and 1922. Simplifying somewhat, there was (probably) a majority in the U.K. as a whole (i.e. the British Isles) for the maintenance of the union; within Ireland (i.e. the whole island) there was a majority in favour of independence for the whole of Ireland; within the six provinces that became Northern Ireland there was a majority for partition as a second best to union; but within two of those six counties there was a majority for unification with the south as a second best to independence for the whole of Ireland.[6]

In cases such as these, the only solution may be to assign rights to the geographically dispersed minority group. The protection of these rights, moreover, could be ensured by constitutional guarantees that make them very hard to abolish. Yet once again we encounter a difficulty. Rights must be *assigned* by someone; they do not fall from heaven. By whom, and by what procedure, ought they to be assigned? One answer has already been suggested. For a right to receive strong constitutional guarantees, it must be strongly supported in the population. If people feel strongly about rights that a majority is not prepared to uphold and if, moreover, the minority is geographically dispersed, emigration and civil war are the only options. And of these, the former may not be available.

The preceding analysis was deliberately brief and incomplete. The intention was neither to present a theory of self-supporting social institutions, nor to explain the endogenous emergence of the variables that the standard theory takes as exogenously given. The suggestions developed above are offered mainly with a view to illustrating one line of foundational research, which proceeds by moving one step further back in the analysis.

[6] Barry (1979, p. 169).

REFERENCES

d'Aspremont, C. and Gevers, L. (1977) 'Equity and the informational basis of collective choice', *Review of Economic Studies* 44, 199–209.

Barry, B. (1979) 'Is democracy special?', in P. Laslett and J. Fishkin (eds.), *Philosophy, Politics and Society*, 5th series, Oxford: Blackwell.

Dasgupta, P., Hammond, P., and Maskin, E. (1979) 'The implementation of social choice rules: some general results on incentive compatibility', *Review of Economic Studies* 46, 185–216.

Gauthier, D. (forthcoming) 'Bargaining and justice'. To appear in *Social Philosophy and Policy*.

Luce, R. D. and Raiffa, H. (1957) *Games and Decisions*, New York: Wiley.

Roberts, K. (1980) 'Social choice theory: the single-profile and multi-profile approaches', *Review of Economic Studies* 47, 441–50.

Sen, A. K. (1970) *Collective Choice and Social Welfare*, San Francisco: Holden Day.

Steedman, I. and Krause, U. (1985) 'Goethe's Faust, Arrow's possibility theorem and the individual decision taker', in J. Elster (ed.), *The Multiple Self*, Cambridge: Cambridge University Press.

1. Lady Chatterley's Lover and Doctor Fischer's Bomb Party: liberalism, Pareto optimality, and the problem of objectionable preferences

BRIAN BARRY

In the ten years since Amartya Sen announced 'the impossibility of a Paretian liberal', the alleged 'liberal paradox' has come in for a good deal of critical discussion.[1] I shall argue in the first section, Lady Chatterley's Lover, that, on a sensible understanding of 'liberalism' and of the significance of something's being 'socially preferred' (in the sense of being picked out by a social welfare function) there is no incompatibility between liberalism and Pareto optimality.[2] In the second section, which is called Doctor Fischer's Bomb Party, I ask on what grounds we might nevertheless disapprove of deals that the parties have a right to make. This is, in my view, the serious question of substantive morality that is left after we have cleared away the problems that are internal to social choice theory. A third concluding section attempts to tie together the two themes.

[1] A. K. Sen (1970a; 1970b, pp. 78–88); an extensive review is provided by Sen (1976).
[2] Let me make it plain at the outset that my object here is not to add yet another 'solution' to the literature (or endorse any that already exists in it). There is no contradiction between saying, 'It would be Pareto-optimal for A to do x', and saying 'A has a right not to do x'. Nothing is therefore required to effect a reconciliation between the two assertions. Similarly, I deny that there is any contradiction between saying that x is a better state of affairs than y and also maintaining that A has a right to bring about y rather than x if he chooses to do so. We should simply reject the suggestion that our moral system has something wrong with it unless it generates the conclusion that if A exercises his right by choosing to do y then it must be a better (or 'socially preferred') state of affairs in which he does y. On the contrary: any view from which it follows that people can never exercise their legitimate rights in ways that have bad results seems to me to stand self-condemned as an absurdity. We know all too well that it happens very often. I do not deny that there is a contradiction between Pareto optimality and Sen's condition L. But, as will be seen, my response is simply that no liberal (or anybody else) has any good reason for holding condition L. In particular, it is quite false that it, or anything like it, is entailed by the proposition that there should be a 'protected sphere' of individual rights. Here again, then, my purpose is not to join in the solution game. My claim is not that condition L needs modification but that it is of an entirely inappropriate kind, forming as it does part of a proposed social welfare function, to correspond to what liberals want to say about the importance of individual rights.

I. Lady Chatterley's Lover

The basic idea of the impossibility result is simple enough. Suppose that we take three conditions that we should like a social welfare function to satisfy: unrestricted domain (U), Pareto optimality (P), and liberalism (L). Then the theorem is that these three requirements, U, P, and L, are incompatible. 'Unrestricted domain' means that we do not place any restrictions on the range of individual preferences entering into the social welfare function. Pareto optimality is the requirement that if everybody prefers one state of affairs to a second, then the first state of affairs is 'socially preferred'; that is to say, ranked higher on the social welfare function. (The question of what exactly is the significance of a social welfare function is one that I shall duck here, not because I want to slight its importance but because I want to take it up at length later.) What the liberal condition means here is that each individual should be 'socially decisive' over some pairs of alternative social outcomes. Candidates for the role of such pairs would be the two states of the world that are identical in every respect except that in one I sleep on my belly and in the other I sleep on my back, or that in one my kitchen walls are painted pink while in the other they are painted crimson. In other words, if I prefer to lie on my belly rather than on my back, the social welfare function should produce the conclusion that, of two states of the world differing only in respect of the way up I lie, the state of the world in which I lie on my belly is socially preferred.

We had better at this point, for the sake of brevity, resort to ps and qs. L specifies that for each i there is a pair of alternatives p and q such that if i prefers p to q, the social welfare function ranks p above q (and conversely if i prefers q to p); and P specifies that, if everyone prefers p to q, the social welfare function ranks p above q. The thesis of the 'impossibility of a Paretian liberal' states that conditions L and P are inconsistent, given U. The proof is very simple and, as far as it goes, incontestable. It runs as follows. Let i be decisive over the pair $\langle p, q \rangle$ and j be decisive over the pair $\langle r, s \rangle$. If i prefers p to q and j prefers r to s, then, by condition L, p is socially preferred to q and r is socially preferred to s. But if i and j both also prefer q to r and s to p, then, by condition P, q is socially preferred to r and s is socially preferred to p. So p is preferred to q, q is preferred to r, r to s, and s to p. This is, obviously, a violation of 'acyclicity', which is normally taken as an absolutely minimal demand to make of a social welfare function.

The entire literature tends to revolve around an example which Sen introduced in 1970 involving the reading of D. H. Lawrence's novel *Lady Chatterley's Lover*. (This of course dates the original treatment fairly precisely, for today stronger meat might be required to make the case plausible.) The example runs as follows. Mr A (the prude) has the following preference order with regard to this supposedly lubricious book: $\bar{a}\bar{b}$ (nobody read it); $a\bar{b}$ (he read it rather than contemplate the lewd Mr B wallowing in that filth); $\bar{a}b$ (Mr B read it); then ab (both read it). Mr B has the following preference order: ab (it would do Mr A good to read it and he would enjoy reading it himself); $a\bar{b}$ (if only one is to read it, let Mr A widen his literary horizons); $\bar{a}b$ (read it himself); $\bar{a}\bar{b}$ (waste of a good book).

It may help to set this out in tabular form. The preferences, then are in the following order:

A *(the prude)*	B *(the lewd)*
$\bar{a}\bar{b}$	ab
$a\bar{b}$	$a\bar{b}$
$\bar{a}b$	$\bar{a}b$
ab	$\bar{a}\bar{b}$

As may immediately be seen, $a\bar{b}$ is Pareto-superior to $\bar{a}b$. That is to say, both A and B prefer it. Yet the liberal principle, according to Sen, demands that the lewd Mr B should read the book and the prudish Mr A should not, since each should decide on his own reading matter. Since I want to maintain that there is no such liberal (or libertarian) principle entailing this conclusion, it would be as well to quote Sen's words:

> On libertarian grounds, it is better that the lewd reads the book rather than nobody, since what the lewd reads is his own business and the lewd does want to read the book; hence $[\bar{a}b]$ is socially better than $[\bar{a}\bar{b}]$. On libertarian grounds again, it is better that nobody reads the book rather than the prude, since whether the prude should read a book or not is his own business, and he does not wish to read the book; hence $[\bar{a}\bar{b}]$ is better than $[a\bar{b}]$.[3]

3 A. K. Sen (1979a, p. 550). Cf. Sen (1976, p. 218).

Thus, we get the 'preference cycle' that constitutes the 'liberal paradox'.

What drives this result is, of course, the fact that both A and B have what Sen calls 'nosy' preferences: A, the prude, is more concerned to prevent B, the lewd, from reading what he regards as a pornographic book than he is to avoid reading it himself (either to save B's soul or because he can't bear to think of him enjoying it); and B is more interested (whether maliciously or in a missionary spirit) to inflict *Lady Chatterley* on A than he is in reading it himself.[4] But is it true that such nosy preferences subvert liberalism? It seems to me that this is not so. Sen claims, as we have seen, that the liberal principle says that it is 'socially better' that the lewd read *Lady Chatterley* and the prude not, but I simply deny there is any such liberal principle. Hence Sen's claim that he has shown 'the incompatibility of the Pareto principle . . . with some relatively mild requirements of personal liberty, for consistent social decisions, given unrestricted domain'[5] simply falls to the ground. For liberal principles do not say in a context like the *Lady Chatterley* case who should read what; rather, liberalism is a doctrine about who should have a right to decide who reads what. Sen, citing Mill and Hayek, suggests that 'considerations of liberty require specification of . . . e.g. whether a particular choice is self-regarding or not . . . , or as falling within a person's "protected sphere" . . .'.[6] Quite so, but let us take note of the purpose for which this information is required. It is brought in at the stage at which we argue about the allocation of rights: we say that when it comes to (say) what they read, people should have power to take their own decisions because this is 'self-regarding' or should be within the 'protected sphere'. But we do not then judge the *use made* of these rights in terms of any principle of

---

[4] Note that it is not merely that A and B have 'nosy preferences', in the sense of caring about each other's reading habits, but that they have *strong* nosy preferences. That is, the 'paradox' would not arise if each had a preference concerning the other's behaviour but, when it came to the crunch, cared more about his own. This would yield the following orderings:

A	B
$\bar{a}\bar{b}$	ab
$\bar{a}b$	$\bar{a}b$
$a\bar{b}$	$a\bar{b}$
ab	$\bar{a}\bar{b}$

Since both prefer $\bar{a}b$ to $a\bar{b}$, conditions L and P both endorse the same outcome here.

[5] A. K. Sen (1979a, p. 549).   [6] *Ibid.*

liberty: the notion that any principle of liberty that would be endorsed by Mill or Hayek is violated by the prude choosing to read *Lady Chatterley* and the lewd not doing so (provided, of course, that they are freely exercising their rights, as in the case stated) is pure fantasy.

Liberalism cannot be connected up to Sen's statement of an alleged liberal principle because it is not a doctrine about what constitutes a 'socially better' state of affairs, where a state is defined by things like '*A* reads *Lady Chatterley* and *B* doesn't'. Rather it is a doctrine about who has what rights to control what. When the liberal principle says that there should be a protected sphere,[7] what that means is that there are some things (e.g. which way up to lie in bed, what colour to paint one's kitchen, or how to spend one's money) about which an individual should be able to decide what to do, without any coercive interference by or on behalf of society. Indeed, further than that, the force of society will be put behind the claim of each individual to do what he likes in such a protected sphere without being subject to coercion by anybody else.

We can express this idea perspicuously by making use of the conceptual apparatus developed by James Coleman.[8] In this schema, actors can control certain events, either individually or jointly. They also have interests in events—not necessarily the same ones as those over which they have control. In the case of a binary choice, an actor's interest in getting the alternative that he prefers may be understood as the difference between the utility that he expects to derive from the favoured alternative and the utility that he expects to derive from the less favoured one. Clearly, as Coleman says, a utility-maximizing actor will seek to 'do two things: first, gain control over those events that interest him, and secondly, exercise that control in such a direction that the outcome he favors occurs'.[9]

And, once we have a structure of rights to control events and of interests in those events:

the principal activities of actors in attempting to realize their interests consists of actions through which they gain effective control of events they are interested in, through giving up effective control

---

[7] A. K. Sen (1976, p. 218).
[8] J. S. Coleman (1973) *The Mathematics of Collective Action*, Ch. 3.   [9] *Ibid.*, p. 73.

over events they are not [or less—B.B.] interested in. In economic activities, this consists of an explicit exchange of control; in other areas of life, including political activities, it consists of less explicit exchange of control often based on informal agreements.[10]

Let us apply these notions to the *Lady Chatterley* case. The initial distribution of control over events provided by a liberal scheme for assigning rights is as follows:

A (the prude)   controls   { *A* read *Lady Chatterley*
                                { *A* not read *Lady Chatterley*

B (the lewd)   controls   { *B* read *Lady Chatterley*
                                { *B* not read *Lady Chatterley*

If each makes his choice independently, *A* will choose not to read *Lady Chatterley* and *B* will choose to do so. However, if we look at the structure of interests, we see that, because both actors have nosy preferences, the structures of rights is out of alignment with the structure of control. Each is more interested in what the other does than in what he does himself.

Ranking of interests by A
1.   *B* not read *Lady Chatterley*
2.   *A* not read *Lady Chatterley*

Ranking of interests by B
1.   *A* read *Lady Chatterley*
2.   *B* read *Lady Chatterley*

It is this structure that makes it mutually advantageous to the two parties to exchange control over the event in which they are less interested for control over the event in which they are more interested. *A* surrenders his right not to read *Lady Chatterley* in return for the more valued control over the reading matter of *B*, and *B* in return surrenders his right to read *Lady Chatterley* in return for the ability, which is more important to him, to have *A* read it.

It is essential to notice that the initial assignment of rights plays a crucial role in bringing about this post-trade outcome in which the prude finishes up reading *Lady Chatterley* and the lewd finishes up not

---

[10] *Ibid.*, p. 75.

reading it.[11] Suppose, for example, that the rights were not distributed so as to give each actor control over his own reading matter, but instead that the prude could determine both what he himself read and also what the lewd read (a lewd minor, perhaps). Then, the outcome that would be brought about by an own-utility-maximizing prude (one, for example, not constrained by considerations of fairness in deciding how to exercise his rights) would be one in which neither reads *Lady Chatterley*.

It should be borne in mind that Pareto optimality is in itself indifferent to distributive considerations. Thus, for example, the outcome in which neither reads *Lady Chatterley's Lover* would be Pareto-optimal, since there is no alternative to it that both parties would prefer. This is the outcome we might expect if the initial assignment of rights allowed the prude to decide on both what he himself reads and what the lewd reads. For clearly, since the lewd has no rights over his reading matter, he has no bargaining counter to offer in order to gain control over events controlled by the prude and move the outcome in a direction he favours. To put it another way, since the prude has all the power, the Pareto-optimal outcome is simply the one that puts *him* as high up his preference rankings as he can get, and to hell with the preferences of the lewd. An exception would arise if the lewd were lucky enough to have his preferences enter into the prude's utility function—but presumably the prude, being a meddlesome prude, does not give weight to others' lewd preferences in arriving at his own preference ordering of outcomes.

The essential point about Pareto optimality, then, is that it simply calls for all mutually advantageous deals to be made, but to pick one Pareto-optimal outcome out of the infinite number of Pareto-optimal outcomes we have to have an initial assignment of rights. The liberal principle is a criterion for the assignment of these rights which tells us who should start with the power to control what events. If we knew the assignment of rights and what Coleman calls the structure of interests, we could, given his own theory, in principle work out the unique

---

[11] Of course, if symmetry were maintained but the initial assignment of rights gave each control over the other's reading matter, there would be no transfer of control (given the nosy preferences that we are postulating) because the structure of control and the structure of interest would then be congruent. However, if each were more interested in what he read than in what the other read, they would transfer control so that the prude did not read *Lady Chatterley* and the lewd did.

Pareto-optimal outcome corresponding to that assignment of rights. In the absence of such a strong theory as Coleman's, the assumption that mutually advantageous exchanges of control over events would be made until no opportunities remained would not generate a unique outcome. There might, in other words, be alternative Pareto-optimal outcomes compatible with a given initial assignment of rights. (We might recall that in the theory of economic bargaining—over an exchange of apples for oranges, say—it requires a very strong theory to pick out a unique point on the contract curve.)

It is important to be clear that there is all the difference in the world between having a right to decide something, where the contents of the decision may be made contingent on what other people offer to make it one way rather than another, and not having a right at all. Only by eliding that distinction can Sen achieve the effect of making us think that there is nothing between totally isolated decision-making in personal matters and a situation in which all preferences, including nosy ones, are thrown into a common pot and some sort of utilitarian calculus applied to them to arrive at politically-enforced decisions about what people must do or must not do (in effect, a set-up in which there are no rights at all).

Thus, in considering how one might come to drop the condition of liberalism, Sen offers this possible argument for doing so: 'The idea that certain things are a person's "personal" affair is insupportable. If the color of Mr. *A*'s walls disturbs Mr. *B*, then it is Mr. *B*'s business as well. If it makes Mr. *A* unhappy that Mr. *B* should lie on his belly while asleep, or that he should read *Lady Chatterly's* [sic] *Lover* while awake, then Mr. *A is* a relevant party to the choice.' He then goes on to say: 'This is, undoubtedly, a possible point of view, and the popularity of rules such as a ban on smoking marijuana, or suppression of homosexual practices or pornography, reflect, at least partly, such a point of view. Public policy is often aimed at imposing on individuals the will of others even in matters that may directly concern only those individuals.'[12] But, he says, to deny the weak liberal condition (which after all requires only that there be *some* alternative over which each person is decisive) is to 'deny even the most limited expressions of individual freedom. And also to deny privacy, since the choice

---

[12] A. K. Sen (1970b, p. 82).

between x and y may be that between being forced to confess on one's personal affairs (x) and not being so forced (y).'[13]

I hope it will be clear by now that this confuses two quite different ideas: that people should never fail to act on their personal preferences in what 'directly concerns' them, and that people should not be *required* to violate their personal preferences in what 'directly concerns' them. The second is, indeed, an authentically liberal idea. But the first is not, as Sen suggests, an essential part of every reasonable conception of liberalism. It might even be said to be antithetical to a conception of liberalism that emphasizes the freedom of individuals to make their own choices with as few constraints as possible. For surely we are more free to choose if we can trade a decision over something we have a right to control in return for control over a decision that we value more, which some other person has a right to take, than we are if some agent of the social welfare function restrains us from doing so. The simple answer, then, is that there is no inconsistency in principle between liberalism and Pareto optimality. Liberalism is, indeed, a principle that picks out a protected sphere, but one that is protected against unwanted interference, not against use in trading with others.

To sum up, if a social welfare function tells us what constitutes a better state of the world, there can be no conflict between any social welfare function, whatever its content, and the principle that there should be a protected sphere within which people shall be legally free to do what they choose. For the two have different subject matters: one is about what is 'socially better', the other about what people shall be able to do without legal coercion.

The only way in which we can create a conflict between a social welfare function and a genuinely liberal principle of this kind is by supposing that what is socially better should be enforced. But then no elaborate argument is needed to establish incompatibility, since that idea is *itself* in conflict with the liberal principle. Thus, again, nothing turns on the particular content of the social welfare function. If the choice between x and y is within A's protected sphere, this entails that A should neither be forced to do x nor be forced to do y.

To ask (on this interpretation of what it implies to say something is socially better) how we should determine, in accordance with liberal-

---

[13] *Ibid.*, p. 83.

ism, whether x or y is socially better is a question obviously doomed to produce an absurd answer, so it is hardly surprising that it is possible to show that any particular answer is absurd. To say 'A's own preference shall determine whether x or y is socially better' is not a 'liberal' answer, because it still presupposes (on this interpretation of what it means to say that something is socially better) that either x or y should be enforced—yet this is exactly what the notion of a protected sphere is designed to deny.

II. Doctor Fischer's Bomb Party

So far in this paper I have sounded, most of the time, like a born-again disciple of Milton Friedman, with the slogan 'Free to Choose' inscribed on my banner. It is now time to dispel that impression. As I said at the outset, I do not find Sen's *Lady Chatterley* example very compelling. But I do share the intuition that seems to lie behind Sen's anti-Paretianism: that some Paretian deals raise problems. However, Sen has misdiagnosed the issue by talking about a 'liberal paradox'. And he has, subsequent to 1970, compounded the error by saying that he prefers 'libertarian' to 'liberal'.[14] For, in the contemporary jargon, a libertarian is precisely somebody who believes that no limits should be set on the drive towards Pareto optimality. The state should not, for example, step in between a willing seller of labour (willing, that is, compared to the alternatives) and a willing buyer by prescribing minimum wage rates. Nor should it prevent people from entering into onerous contracts (e.g. at usurious rates of interest) if they regard that as advantageous to themselves. Thus, it is clear that libertarianism, so conceived, is in no sense incompatible with Pareto optimality. Rather, Pareto optimality is a natural part of the kind of social philosophy (in effect an ideology constructed out of outdated introductory economics textbooks) known as libertarianism.

This ideology assumes, in effect, that if people are to have a right in general to do what they choose in some area (e.g. buy or sell labour, lend or borrow money), it is better for there to be no legal restrictions on the exercise of the right. But there are in some cases good reasons for setting limits to the range of transactions that should be permitted. Minimum wage laws and usury laws can operate so as to strengthen the

14 A. K. Sen (1976, p. 218).

position of the weak, providing by legal enactment the equivalent of an agreement that the weaker parties would have found it advantageous to conclude among themselves, if the problem of collective action did not arise. On somewhat similar lines, legal systems characteristically do not allow the consent of the victim to count as a defence to a charge of inflicting bodily harm or death. The rationale is, plausibly, that the position of potential victims would be weakened if they could validly consent. The prohibition of contracts of servitude and of debt peonage can be justified in the same way. These limitations do, of course, prevent the achievement of Pareto optimality, if we take a regime of unrestricted rights as our baseline and look at potential transactions one by one. But it is also inconsistent with libertarianism, as I have defined it.

It is, then, quite reasonable to assign rights in society so as to take account of the strategic relations of the parties. In the cases I gave, one side actually has an interest in being prevented from entering into certain kinds of legally effective agreements, thus illustrating, *contra* Friedman, that it is not always advantageous to be 'free to choose'. Although in general people can do better for themselves (i.e. achieve more preferred outcomes) by being legally free to deploy their rights in any way they wish, there are some cases where they can achieve a more preferred outcome if their rights are restricted so that they are prevented from giving up their rights or from exchanging them on particular terms.

At this point, however, we may begin to wonder whether the notion of Pareto optimality is well defined in such contexts. I said that limits on the exercise of a right prevent the achievement of Pareto optimality if we take as a baseline the absence of restrictions and if we consider the potential transactions one by one, independently of one another. But why should we? What we have here is a situation in which someone may be better off doing a certain deal than not doing it if he is permitted to do it, but would be better off still if he were not permitted to do it. Since *ex hypothesi* his position in the second rights regime is preferable to any open to him (either by dealing or not dealing) in the first rights regime, we cannot say that either of the regimes is Pareto-superior to the other. We have, rather, two regimes with different distributions among the actors, neither of which dominates the other. On this line of reasoning, we can say that both regimes have the possibility within them of attaining (different) Pareto-optimal outcomes.

Fortunately, conundrums of this kind do not have to be straightened out before I can present the problem to which this section of the paper is devoted. Let us suppose that we have some system of rights established, and that it is of a conventional liberal nature. That is to say, although it restricts the exercise of rights along some of the lines just mentioned, it leaves open a wide range of discretionary behaviour. In addition to the resources for exchange offered by the market, then, we will suppose that people have many other opportunities to trade rights over things they control for things that other people control.

The question I want to raise is the following. When people operate within this established framework of rights, are there Pareto-optimal deals which they can properly be condemned for making? We are thus looking at cases where someone uses a right he unquestionably has (e.g. to read or not read *Lady Chatterley*) in order to try to influence, by making a contingent offer to act one way rather than another, the use another person makes of a right that *he* unquestionably has (e.g. to read or not read *Lady Chatterley*). Interestingly enough, Sen himself, in an article written three years after the survey article from which I was quoting before, broke away far enough from the social choice terminology to present the question in these terms, though without repudiating the implicit claim that what he said could be translated into the language of social welfare functions. He said there that a trade between the lewd and the prude:

> raises a deeper question, viz., whether *having a right* based on the 'personal' nature of some decisions (in this case the right to read what one likes and shun what one does not wish to read) must invariably imply being free *to trade that right* for some other gain, irrespective of the nature of the gain (in this case the lewd's gain consists in getting pleasure from the prude's discomfiture, and the prude's gain in avoiding the discomfort of knowing that the lewd is reading a book that he—the prude—disapproves of). If the answer to this question is yes, then clearly the criticism of the Pareto principle would not apply to this case. I believe it is possible to question such an affirmative answer, but I resist the temptation to go further into this complex issue. . . .[15]

[15] A. K. Sen (1979a, p. 552).

It is precisely that complex issue that I wish to address here.

To begin with, let us get it quite clear that there is nothing wrong in general with attempts to manipulate one's use of rights to advance nosy preferences. Suppose that I have a purely disinterested nosy preference for your giving up smoking (it isn't that the smoke bothers me, etc.) and you have likewise a preference for my losing weight based on a disinterested concern for my health rather than a desire to look at someone trimmer. It seems hard to see how any moralist could object to our making a deal in order to further our nosy ends. There is, surely, a version of the *Lady Chatterley* case that comes close to this, where the contingent offers are well-intentioned in a somewhat parallel way. It is only by attributing nasty motives to the lewd (making the prude squirm) and reducing the prude's concern for the lewd to the self-centred one of avoiding his own discomfort at knowing the lewd is reading the book that Sen is able to make us feel any qualms at all about the propriety of a trade. Even then, the case is one that I find hard to take seriously. There is no question that the lewd and the prude of Sen's depiction would be better people if they had different preferences but, given the preferences they have, I still feel reluctant to say that they shouldn't act to further them. In fact, it seems to me that they deserve one another.

Let me take up instead the second novel that I have included in my title, Graham Greene's *Doctor Fischer of Geneva or the Bomb Party*.[16] The eponymous Dr Fischer is a very rich Swiss who, for reasons that we need not enter into here, wishes to have his contempt for the human race reinforced and justified. To this end he assembles a group of wealthy toadies (or Toads, as his daughter calls them) who endure all kinds of insults and humiliations at the hands of Dr Fischer in return for lavish gifts. Although they could, as Dr Fischer frequently emphasizes, quite well afford to buy the same things for themselves, they are too mean to spend their own money and greedy enough to put up with a lot of indignities in order to get them free. The climactic 'bomb party'[17] is unsuitable for our purposes because I suspect that, even under Swiss law, consent is no defence to a charge of bodily harm, and Dr Fischer by offering his guests Christmas crackers which have either a large cheque or a small bomb in them, and inviting them to pull one each and take the consequences, is on shaky legal ground. And even if he isn't legally guilty, we may still wish to hold the right to

[16] New York: Simon and Schuster, 1980.   [17] *Ibid.*, pp. 65–6.

bodily integrity to be morally inalienable. Let us therefore take the earlier dinner party which the narrator (who is married to Dr Fischer's daughter) attends, since this raises no problems of violating the rights of the guests. At this party, Dr Fischer dines on caviare, while his guests are offered nothing but cold porridge. As Dr Fischer points out, they are perfectly free not to eat it if they choose not to. As the narrator recounts it to his wife afterwards:

> 'Mrs. Montgomery [one of the Toads] said I should have been sent from the table as soon as I refused to eat the porridge. "Any of you could have done the same," your father said. "Then what would you have done with all the presents?" she asked. "Perhaps I would have doubled the stakes next time," he said.'
> 'Stakes? What did he mean?'
> 'I suppose he meant his bet on their greed against their humiliation.'[18]

I shall take it as uncontroversial that Dr Fischer would have been a better man if he had not had the desires that led him to give such dinner parties. But can we say that Dr Fischer ought, morally, not to have acted as he did? Can we say that it was not merely bad-producing conduct (like painting one's kitchen in clashing colours) but wrong?

I do myself feel strongly inclined to say this. If I try to determine what special features of the case make me want to say that it was wrong for Dr Fischer to act on his objectionable preferences, I seem to find two. The first is that this is a case of playing on a weakness of character—in this case greed—for one's own satisfaction. The other, somewhat related point is that Dr Fischer is deliberately corrupting the Toads, that is to say, making their characters worse in respect of the flaw of excessive greed. When Dr Fischer's invitation to the narrator to come to the dinner party arrives, his wife says: 'He wants you to join the Toads.' The following dialogue ensues: 'But I've nothing against the Toads. Are they really as bad as you say? . . .' 'They weren't always Toads, I suppose. He's corrupted all of them.'[19]

Dr Fischer played on human weakness. Another type of case is that where what is taken advantage of is economic weakness. Thus, where

---

[18] *Ibid.*, pp. 73–4.   [19] *Ibid.*, pp. 40–1.

there are great disparities in economic circumstances between the parties, and in particular where one is in desperate need, the other (better-off) party may take advantage of the economically weaker one. Note that the resultant deal is certainly Pareto-optimal in that the weaker party prefers the deal to the *status quo ante*. Our criticism is of the stronger party for offering those terms. The underlying idea is again that there are ways in which people shouldn't treat others.

Examples of this second kind of case would be as follows:

(a) Offers that express contempt, by emphasizing the worthlessness of the person. An example would be paying people to dig holes and fill them up again. We might contrast this with the superficially similar case of a landlord paying men (as sometimes happened in earlier centuries) to carry out ornamental work on his estate purely in order to provide employment for them. In such a case, one may well object to the distribution of initial resources that puts the parties in the position where that is a Pareto-optimal deal but not, it seems to me, to the transaction itself.

(b) Offers that are intrinsically degrading—to do such things as eating excrement, blaspheming against one's religion, prostitution. (This is, of course, on the analysis that prostitution is taking advantage of poor alternative employment for women. Note, however, that this analysis can be used either to criticize prostitution or to criticize the economic position of women. In other words, if you can't convince people that working in a match factory or a sewing room for the available pay is bad, you may still be able to convince them by pointing out that many girls prefer prostitution. This was, of course, Shaw's strategy in *Mrs Warren's Profession*.)

A third case of offers that ought not to be made is, I would suggest, those that are really coercive offers even though they are not caught within the usual (or legally enforcible) conception. In effect, these involve not appealing to psychological or economic weakness but to the recipient's better nature, but in an inappropriate way. Examples would be: 'If you don't marry me, I'll kill myself'; 'If you don't give me the job, it will blight my life'; 'Only by doing what I want can you save me from committing a mortal sin.' The point of these cases is that the person is not making a threat ruled out by the usual bans on coercion. Yet at the same time there is no doubt that the other person's decision is taken under duress. 'If you don't do it, I'll kill myself' may well be much more really coercive than 'If you don't do it, I'll punch you on the nose.'

To sum up, the question has been one of actions taken to encourage someone to do what you want him to, taken against a background where he is free to choose to accept or reject whatever offer you make. I have suggested that everything turns on that phrase 'free to choose': being free to choose is not simply equivalent to not being coerced in the sense of threatened with the deprivation of a right. It has got to include the idea that not going along with the deal that is offered is an acceptable state of affairs. The second and third types of exception I gave are both ways in which this condition fails: in the third you are going to be made to feel bad if you don't accept (so it may be treated as an implicit sanction) and in the second you are not really free to reject it because the *status quo* is so unbearable that you have to take anything that promises to alleviate it. (A subset that gets closer to implicit coercion is where in the absence of a deal things get worse and worse, not because of the other's action but simply in the nature of the case—the obvious examples being where you are running out of money without a job, running out of water in the desert, etc.) The Dr Fischer case fits in more awkwardly, but we might say that weakness of will in the face of temptation is a lack of freedom to choose. In the absence of such conditions—so that the person to whom the offer is made is genuinely free to refuse it—is there any reason why, *given* the preferences, it is wrong to try to change the incentives facing somebody?

I do not think so. When it comes down to it, the only really objectionable cases do seem to be the three kinds of case I have picked out already. And the only plausible way of getting the *Lady Chatterley* case in is by imagining that the prude is presented with the offer as a challenge to his integrity. Suppose that the lewd says 'You always say that it matters a lot to you that people don't degrade themselves by reading filth. Here's a chance to stop me doing so, and the only cost to you is reading it yourself—and you're so incorruptible it won't matter to you.'

Examples of objectionable attempts to get compliance with nosy preferences generally rely on a background of inequality. For example, the effort of the Victorian middle class to regulate the lives of their servants over and above what was defined in doing the job for which they were employed is no doubt revolting to any reasonable person. But isn't this because their market position was such that they could impose these conditions as the norm? Suppose that a couple now wanted to employ a live-in nanny for their children and insisted on no visits from boy friends. In the present market this would presumably cost quite a premium. But then if somebody thinks it is worth the premium and takes the job on those conditions do we have any reason to think that that is a bad arrangement? Given a correct distribution of rights, then, my conclusion is that we can afford to be pretty tolerant of efforts to further nosy preferences, even unworthy ones.

III. Discussion

This paper has raised two questions. The first is whether there is some inherent conflict between the Pareto principle and some liberal principle of a kind that would be endorsed by J. S. Mill or F. A. von Hayek. The second is whether we have grounds for the moral condemnation of the use people sometimes make of the area of discretion granted them by a system of rights in offering to act in a certain way in return for some change in behaviour by another. This may be paraphrased as a question whether Pareto-optimal moves can be undesirable. The answer to the first question is a clearcut 'no': there is no such contradiction. The answer to the second is a more hesitant 'yes': under some circumstances mutually preferred actions may be open to censure.

The key to the first question is that the Pareto principle and the liberal principle have different subject-matters, so they cannot conflict directly. The Pareto principle is a criterion for judging the goodness or badness of states of affairs, whereas the liberal principle à la Mill or Hayek is a criterion for assigning rights to individuals (and perhaps collectivities).

To spell this out a little more fully, here is a statement by Leontief of the Pareto principle, as quoted by Sen: 'The social welfare is increased whenever at least one of the individual utilities on which it depends is raised while none is reduced.'[20] And, in Sen's translation into his own lingo: 'For all pairs of states x, y, . . . if everyone has at least as much utility in x as in y, and someone has more utility in x than in y, then x is

[20] A. K. Sen (1979a, p. 537).

socially better than *y*.'[21] Note that this is a statement about what makes social welfare greater or makes one state of affairs socially better than another. It says nothing about what rights individuals should have or indeed what use they can, with propriety, make of those rights.

Now recall Mill's famous 'simple principle', which is, I suppose, a paradigmatic statement of the kind of liberal principle that is supposed to conflict with the Pareto principle:

> The object of this Essay is to assert one very simple principle, as entitled to govern absolutely the dealings of society with the individual in the way of compulsion and control, whether the means used be physical force in the form of legal penalties, or the moral coercion of public opinion. That principle is, that the sole end for which mankind is warranted, individually or collectively, in interfering with the liberty of action of any of their number, is self-protection. That the only purpose for which power can be rightfully exercised over any member of a civilized community, against his will, is to prevent harm to others. His own good, either physical or moral, is not a sufficient warrant. He cannot rightfully be compelled to do or forbear because it will be better for him to do so, because it will make him happier, because, in the opinions of others, to do so would be wise, or even right. These are good reasons for remonstrating with him, or reasoning with him, or persuading him, or entreating him, but not for compelling him, or visiting him with any evil in case he do otherwise.[22]

Not only does this not conflict with principles for the evaluation of states of affairs but it presupposes them. For Mill's whole point is that an individual's own good might be increased by his doing something where nobody else is adversely affected, and that this is nevertheless no ground for 'interfering' by means of legal or social sanctions. (Note, however, that 'remonstrance' is not a sanction.) It is thus historically bizarre to suppose that Mill would be fazed by any 'dilemma of a Paretian liberal', since he quite explicitly dissociated his 'simple principle' from any judgment of consequences. He accepts that an individual may, in exercising his rights, pass up a chance to make himself better off and nobody else worse off; but he denies that that is

[21] *Ibid.*, p. 538).   [22] J. S. Mill (1859).

any reason for the rest of us to make him do it if he chooses not to.

There is no need for present purposes to take up the hoary question whether, if the run is sufficiently long (e.g. measured in centuries) and the definition of utility sufficiently elastic (e.g. 'the permanent interests of man as a progressive being'), Mill's 'simple principle' is compatible with a very indirect sort of utilitarianism. The point that matters here is that a conflict between a liberal system of rights and the Pareto principle can arise only if (a) every single act is to be assessed according to the Pareto principle, and if (b) enforcement is considered to be an appropriate response to infractions of it. And it is made quite clear by Mill that this is exactly what he wants to deny; in fact the 'simple principle' might be said to consist of the denial.

It is, then, undeniable that if we propose a criterion for a good state of affairs like Pareto optimality, or, more full-bloodedly, maximizing aggregate utility, and if we also say that legislators and government officials should seek on a case-by-case basis to enforce the pursuit of this end, then farewell legal rights. And if we also assign a duty to everybody to coordinate social sanctions in order to maintain pressure towards the pursuit of the collective end, we have Mill's nightmare (how quaint it seems now!) of an excessive weight of public opinion on individual freedom. But what must be emphasized is that there is absolutely no inconsistency between holding that it would be a better state of the world if someone refrained from doing x (because doing it will be 'physically' or 'morally'—Mill's terms—deleterious to him with no offsetting gains to others) and at the same time holding that he should not be legally or socially coerced into refraining from doing x.

It should be observed that I have been talking about the absence of enforcement (legal and, in Mill's extension, social) in certain areas as the essence of liberalism, and this seems to me historically correct. There is no inconsistency between liberalism and the view that, as a matter of individual morality, each person has a duty to maximize the amount of good in the universe. Thus, Godwin, in *Political Justice*, is archetypically liberal (even anticipating Mill's extension of the doctrine from law to public opinion) in looking forward to the end of all legal and social sanctions, while at the same time putting forward a fanatically rigorous doctrine of individual morality according to which every shilling in my pocket has its destined recipient, namely the person in all the world who will derive the most happiness from spending it.[23] This kind of strict consequentialist doctrine has few adherents as a theory and probably none in practice, but there is nothing anti-liberal about it. It can be combined, as it was in Godwin's case, with a strong endorsement of individual rights against coercion. Indeed, there is much to be said for Godwin's view that the relaxation of legal and social controls can be tolerated only when it occurs in a society within which stringent standards of individual morality are widespread. These standards need not be strictly consequentialist but they must acknowledge, in Justice Frankfurter's words, that 'much that is legally permitted is repugnant to the civilized mind'.[24]

To sum up, we can say that there are (at least) three things to be distinguished: what constitutes a good state of the world (this is what a social welfare function [SWF] should tell us); what rights individuals should be granted, that is to say in what areas of conduct they should be free from legal and (if we follow Mill) social coercion; and what, morally speaking, individuals have a duty to do. ('Duty' is of course just one out of the family of terms used in moral judgment, which are not interchangeable. For the present strictly limited purpose, however, it can stand in for the rest.) How these three are related is a question, or set of questions, in moral and political philosophy. It is a substantive issue and is not to be settled by consulting the meanings of words. (Moore, for example, was simply wrong in regarding the proposition that one has a duty to maximize the amount of good in the universe as an analytic truth.) This is not the place to discuss the substantive issue. The point that has to be made here is that the *only* way in which a conflict arises between liberalism and Pareto optimality is if we adopt the meta-principle that not only do individuals have a moral duty to pursue the good (i.e. implement the SWF), but that 'mankind, individually or collectively' is warranted in *enforcing* the SWF using legal or social sanctions. But, to repeat, there is nothing in the notion of SWF that entails that what is 'socially better' must be (or may be) brought about by coercion.

Suppose that my 'personal' preference (i.e. consulting nobody but myself) is to have a crimson kitchen or read *Lady Chatterley's Lover*. And now suppose that somebody with nosy preferences makes me an offer to change my mind, e.g. offers me $1,000 to switch to pink (thus appealing to my preference for more rather than less disposable

---

[23] W. Godwin (1971).   [24] Quoted by Gerald Grant (1981, p. 141).

income) or offers to read *Lady Chatterley* if I don't (thus appealing to *my* nosy preference in regard to his reading matter). And suppose that the offer is, given my overall preference scheme, attractive: I was almost indifferent between crimson and pink anyway, and would actually have been happy to change for $10; I was only mildly keen on reading *Lady Chatterley* but attach great importance to the other's reading it. Then clearly Pareto optimality (which does here seem to be no more than an expression of elementary rationality) says that it is a better state of the world if I accept the offer: both parties gain and nobody else is affected either way. But nothing immediately follows from this about social intervention. Unless we adopt the additional principle that the SWF is to be enforced, I still have the right to paint my kitchen any colour I like or read whatever books I like. If I choose to act like a damn fool and turn down a deal that would move me up my preference ordering, the fact that I am also failing to move the outcome up a notch on the SWF does not mean that anybody can call a policeman and make me consummate the deal.

It is essential to distinguish between, on the one hand, an actual system—a constitution and a set of laws—that specifies who can control what events and, on the other hand, a criterion on the basis of which one might judge that a certain state of society is morally preferable to some alternative state of society. A SWF is a proposal for a criterion that will tell us what is a better state of society. Sen, by talking about 'social decision functions' and 'collective choice rules' as well as 'social welfare functions', makes it sound, to the unwary reader, as if the topic under discussion were actual constitutional rules. But it is not. It is still simply ideas people might have about what are better or worse states of affairs.

Now it is, of course, true that an interesting theoretical question arises if I support some decision-making rule, that is to say some actual system for the allocation of power, and disapprove of what comes out of it. For example, I may think that the majority has (via some representative system) a right to decide what is done about something and regard the majority decision as unjust. Or I may think that you quite properly have the right to the last word about the colour of your kitchen walls but also think that your decision is deplorable. Undeniably, I may say, you should have the right to decide on their colour. But why should that choice be beyond criticism on the basis of some reasonable criteria for good outcomes? Pareto optimality is by no means the only criterion for criticizing the choice of colour scheme. It might be said that, although you are quite within your rights in having any colours you choose in your kitchen, it would have been better if you had taken account of the sensibilities of others, or shown a less vulgar taste, or spent the money on something other than redecoration, or any of a thousand things. Liberalism says that nobody can make you paint your kitchen any colour you don't want it painted. It does not say that everybody else has to like it, or agree that your choice is the 'socially best' one.

The point is in fact a quite general one that applies either to (real) social decision procedures or to individual rights. In the first case it spawns the so-called 'paradox in the theory of democracy' (not to be confused with the 'paradox of voting') that Richard Wollheim claimed to discover.[25] In the second case it gives us the puzzle how there can be a 'right to do wrong'—for a right only to do right would be no right at all.[26] However, we do not in practice find any great difficulty in voting one way and still accepting that, if we were in the minority, the actual decision should go with the majority. Similarly, we do not in practice (most of us, anyway) feel that there is really any contradiction in saying that somebody should have a right to do either x or y but that what he or she really ought to do is x—even if y is the choice made by the right-holder. And in my view we are quite correct in this.[27]

It must be acknowledged that if one believes that each individual has a duty to maximize the good, and if the good (or SWF) includes Pareto optimality among its criteria, there will be a moral duty to pursue Pareto optimality wherever it leads. This clearly would entail that, if nobody else was affected, you would have a duty to get as high up your own preference rankings as possible. And it would mean that, since your all-things-considered highest preference on almost anything could always be changed by a big enough contingent offer, you could be manipulated into finding that almost anything might be your duty.

Of course, problems arise here about the meaning of 'preference'. If we say that the desire not to be manipulated is itself a wish that can enter into an all-things-considered preference, you may be able to say

[25] R. Wollheim (1962).   [26] See J. Waldron (1981).
[27] I have tried to make this clear in 'Wollheim's paradox: comment' (1973).

that you *prefer* not to accept a fantastic offer. But then preferences and actions become analytically indistinguishable. If we retain a more normal concept of preference (of the kind that Sen himself requires) so that we can say that something is preferable, on my scale of preferences, but I nevertheless choose not to do it, we will have to conclude that a duty to pursue Pareto optimality will be a moral constraint on what might reasonably be thought of as an area of individual discretion. The solution if we don't like this is to reject the premise that gives the trouble: the idea that each person has at every moment a duty to pursue the good (including his or her own good where that of others does not enter into competition with it).

One possible way of accounting for Sen's invention of condition L—but I must insist that this is speculative—is that it is an attempt to incorporate a sphere of moral indifference into a strictly consequentialist moral system. There is some evidence for this interpretation in that he began a relatively recent article by saying that in it, 'without disputing the acceptability of consequentialism', he intended, among other things, to argue that Paretianism 'deserves rejection in its general form'.[28] And one way of interpreting the driving force behind condition L is that there ought to be not merely a protected sphere in the Mill/Hayek sense of one immune to legal or social coercion, but one in which what I do is *morally* my own business, in other words where I have no duty to do one thing *or* the other. If this is the intention lying behind condition L then it seems to me misguided. For there is a much more direct way of getting rid of the problem of a vanishingly small sphere of moral indifference if that is the worry, namely abandoning strict consequentialism. And Sen's attempt to incorporate a condition guaranteeing a sphere of moral indifference into a consequentialist morality has the unfortunate effect of undermining the judgments of outcomes that we should actually be prepared to make.

We can see how this happens by following Sen's development of condition L. His informal statement of the principle of 'personal liberty' runs as follows: 'There are certain personal matters in which a person should be free to decide what should happen, and in choices over these things whatever he or she thinks is better must be taken to be better for the society as a whole, no matter what others think.'[29] Now the first half of this is fair enough. And it is precisely what is guaranteed

[28] A. K. Sen (1979b, pp. 463–4).   [29] A. K. Sen (1976, p. 217).

by a set of liberal rights. The second half could scarcely be said (as Sen claims) to be one of the 'more widely used principles in evaluating social states',[30] since it is not even intelligible as it stands. The assertion has to be interpreted within the framework of social choice theory.

'Socially better' means 'picked out by the social welfare function', and condition L means that over certain pairs of choices (at least one pair) each person should be decisive in determining what is picked out by the SWF. Clearly, the question to be raised here is: why should it be supposed that a liberal (or indeed anyone else) would wish to endorse Sen's condition L? Why should I wish to insist not only that I should be able actually to decide what happens, e.g. whether my kitchen walls are pink or crimson, but also whether or not it will be 'socially better' that my walls be pink rather than crimson? On the face of it, the latter has less to do with individual autonomy than with megalomania. Surely it ought to be enough that I can decide what colour my kitchen walls are. To ask to be 'socially decisive' in the recondite sense given to that term by social choice theory seems rather presumptuous.

Why would we want to make such a claim? The only explanation I can think of is that, given consequentialism, this is a way of squeezing in an area of moral indifference. But the trouble with this is that, while we may not want to insist that people have a *duty* to (say) act in accordance with the Pareto principle, we may often want to say that it would be a better state of the world if they did, and this possibility is denied by condition L.

Thus, Sen has written that it could be that:

> while I would prefer you to read what I consider to be good literature as opposed to what appears to me to be muck, I do not want my preferences to count in the social evaluation as to whether it is better that you read good literature or bury yourself in muck. I might accept taste differences as legitimate and accept the greater relevance of your taste in matters that I agree are essentially your 'concern'.[31]

But the notion that what you read is your business does not entail *de gustibus non disputandum*. Your taste should, if you insist on following it, have not merely greater but exclusive relevance to the question of

---

[30] *Ibid.*   [31] *Ibid.*, p. 236.

what you actually do. But surely I must, if I believe that I have any judgments of taste about literature or colour schemes (as against merely 'knowing what I like'), apply *my* standards to your reading matter or your kitchen walls. If I believe that *Lady Chatterley* is muck or crimson walls are inappropriate to a kitchen, then that's what I think. I can't be expected to change my mind merely because you happen to think otherwise.

Perhaps the underlying problem here is that the word 'preference' is being grossly overworked. Part of the logical positivist baggage that orthodox welfare economics carries around with it is the idea that judgments of better and worse have no cognitive content but are simply expressions of attitudes. Hence, they can be assimilated quite properly to other forms of preference. When we combine this with the other dogma derived (incorrectly) from crude verificationism to the effect that 'interpersonal comparisons of utility' are 'meaningless' (or themselves express a 'value judgment' construed as something without cognitive content), we get the characteristic form of post-Arrow welfare economics in which 'social welfare' is derived from some process of aggregating preferences.

Sen wants to challenge this kind of 'welfarism', but he is insufficiently iconoclastic. The conclusion he wishes to draw from the Paretian liberal dilemma is that we need more than utility information (i.e. in this case preference information) in order to assess social welfare. We need, for example, to know the source of the preference: why does the lewd prefer that the prude read *Lady Chatterley's Lover*? But what he does not challenge is the idea that moral or aesthetic judgments should be treated as preferences and put into a SWF. The only question he raises is whether all preferences should 'count in determining social choice'.[32] So, he implies, nosy (or at any rate nastily nosy) preferences should count less than 'personal' ones, or perhaps they should not count at all.

This, however, is an attempt to stop the mischief at too late a point. The move that should be blocked is the one that treats moral or aesthetic judgments as preferences to be cranked into a SWF along with other preferences. If I believe that *Lady Chatterley's Lover* is muck or that crimson walls are unsuitable for a kitchen, then I presumably think it would be a better state of the world if you didn't read *Lady*

[32] *Ibid.*

Chatterley or paint your kitchen walls crimson. But the fact that I believe that is not *itself* what makes it better—nor does it contribute even a tiny bit to making it better. If we add strict consequentialism, so that you have a duty to go for the best state of affairs possible, it will follow from my beliefs that you have a moral duty not to read *Lady Chatterley* or paint your kitchen walls crimson. But the ground for the duty is that (I believe) doing those things would bring about an inferior state of affairs to some other that it is open to you to bring about. The ground is *not* that I would 'prefer' you to refrain—or that I and many other judges of literature or colour schemes share the same 'preference'. I may not, indeed, have any preference in the matter at all, in any ordinary sense of the term. I may simply have formed a quite disinterested aesthetic judgment on the question of *Lady Chatterley* or crimson decor. I may not care a bit what you do, but that obviously doesn't give me any reason for withdrawing my judgment.

We might do better, if we are going to assimilate moral and aesthetic judgments to any other kind of thing, to assimilate them to beliefs about matters of fact. If I believe that the Sears Tower in Chicago is taller than the Eiffel Tower, we presumably are under no temptation to turn that into a statement of a preference. And if I have the (fairly incontrovertible) view that it is better that people believe what is true than what is false, it will follow that I think it would be better for people to believe that the Sears Tower is taller than the Eiffel Tower. But surely it should be plain here that they should (in my view) believe it because it is (in my belief) true—not to please me by falling in with my 'preference in the matter'.

Thus, the belief that it would be a better state of the world if you didn't read *Lady Chatterley's Lover* is itself a social welfare judgment, which I arrive at after taking account of the fact that you would like to read the book (and giving that whatever weight I think appropriate), the effects I think the book would have on you, and anything else that seems to me relevant. My belief is just that, and there is no reason on earth why anybody else (or I myself) should treat it as a preference and put it in with other preferences to determine what would advance 'social welfare'. I have already, *ex hypothesi*, taken account of everything I consider relevant in arriving at my judgment about where the social welfare lies. If somebody else were trying to form a judgment on this point, he too would have to form an opinion of what the book would do to you, find out if you wanted to read it, and so on. But he would be rather eccentric to take account of *my* judgment as an ingredient in *his* judgment about social welfare. (He might, of course, regard me as a literary, or moral, authority, but then he would follow my judgment in reaching his own, not incorporate it as a 'preference'.)

If Sen's concern is with the construction of SWFs out of ill-assorted materials, it is entirely well placed. Indeed, I hope it will not be too self-indulgent to point out that I said this myself, five years before Sen published his Paretian liberal paradox, in my *Political Argument*. In discussing majoritarianism—the idea that things should be done if a majority wants them done—I suggested that we should, if we apply the principle at all, count only privately-oriented wants:

> The justification for counting only privately-oriented wants is that it avoids at one stroke the most objectionable feature of the majoritarian principle, namely the way in which it commits one to handing over questions of right and wrong, justice and injustice, to the majority of a group in which one's own voice counts only as one. Yet at the same time the amendment leaves one free to take account of desires which are put forward simply by people as wants in matters affecting themselves. This may seem high-handed at first sight, but further reflection suggests that what I have called 'publicly-oriented wants' are not actually put forward as wants at all. To treat them as wants is to degrade them and to fall into absurdity.
>
> Suppose that I am making up my mind whether it is fair for the *A*'s to get more of something than the *B*'s; and the *A*'s and *B*'s are the only people directly affected by the division, in the sense of 'affected' which I have defined. Should I, in making up my mind, take account of the opinions of the *C*'s in the matter? I may, of course, let them weigh with me as having a certain authority, but surely it would be ridiculous to mix in the wants of the *C*'s for, say, the *A*'s to win, consequential on their belief that the *A*'s have the best case, on an equal footing with the privately-oriented wants of the *A*'s and *B*'s.[33]

The relevance of this for social choice theory is fairly radical, because it entails that we should reject the whole idea of aggregating preferences—including individuals' judgments about social welfare—into

[33] Barry (1965).

some sort of *social* SWF. The error of supposing that this is a sensible thing to try to do goes back to Kenneth Arrow's (1951) book *Social Choice and Individual Values* (though, as far as I can tell, no further). And a devastating criticism of it was made as early as the next year by Ian Little, who wrote:

> Arrow calls his function both a social welfare function and a decision-making process. He believes that 'one of the great advantages of abstract postulational methods is the fact that the same system may be given different interpretations permitting a considerable saving of time'. Yes, but we must be careful not to give such a system a nonsensical interpretation, and it will be my contention that to interpret it as a social welfare function *is* to give a nonsensical interpretation.
>
> Imagine the system as a machine which produces a card on which it is written '*x* is better than *y*,' or vice versa, when all individual answers to the question 'Is *x* better than *y*?' have been fed into it. What significance are we to attach to the sentence on the card, i.e. to the resulting 'master' order? First, it is clear that the sentence, although it is a sentence employing ethical terms, is not a value *judgement*. Every value judgement must be *someone's* judgement of values. If there are *n* people filling in cards to be fed into the machine, then we have *n* value judgements, not *n* + 1. The sentence which the machine produces expresses a ruling, or decision, which is different in kind from what is expressed by the sentences fed into it. The latter express value judgements; the former expresses a ruling between these judgements. Thus we can legitimately call the machine, or function, a decision-making process.
>
> But what would it mean to call the machine a social welfare function? One would be asserting, in effect, that if the machine decided in favour of *x* rather than in favour of *y*, then *x* would produce more social welfare than *y* or simply be more desirable than *y*. This is clearly a value judgement, but it is, of course, a value judgement made by the person who calls the machine a social welfare function.[34]

There is only one point on which I would like to modify what Little

[34] I. M. D. Little (1952).

said. I do not think that it makes any more sense to say that individual judgments are aggregated into a ruling or decision than it does to say that they are aggregated into a social welfare judgment. What goes into a social decision (that is to say, a real decision with real effects) are votes, not expressions of views about social welfare. What comes out (say of referendums) are decisions not on where the social welfare lies, but on what is to be done. To tie this together with the earlier discussion of (legal) rights, we may say that rights and votes go together in one box (the box marked 'control over actions') while judgments of social welfare go into another box. And this box does not in addition contain anything corresponding to a vote tally. There simply is no such thing as a social welfare judgment compounded out of social welfare judgments. The thing is a logical monstrosity. As Little said, if you feel you should take account of other people's judgments of social welfare, not merely as suggesting where the truth lies but as arguments in your own 'social welfare function', treating them as 'preferences', that is logically possible—though, I would immediately wish to add, morally obtuse. But then the result of that process is *your* judgment about where the social welfare lies. It is not some sort of superordinate judgment that is nobody's judgment and everybody's judgment at the same time.

It may then be asked 'But what happens if the people in a society reach different conclusions about the social welfare?' The answer is, of course: what happens now. Different people do, of course, disagree about what would make the world better. They argue about it, but there is no guarantee that they will finish up by agreeing. What we neither have nor could have is some algorithm for taking these divergent judgments and producing some 'social' judgment.

What we do need, in order to have a stable society, is a constitution that specifies how collectively binding *decisions* are to be taken. And if the society is to be liberal as well as stable, either the constitution or the legislation enacted under it should set out individual rights. But this is, again, in the sphere of control, not the sphere of judgment.

These rights, since they specify what actions people will be allowed to take without being exposed to legal or (in Mill's extension) social sanctions, have no direct connection with anybody's judgments about what makes a better or worse state of affairs. What Sen offers as his liberal principle—condition L—has no connection with rights, and thus no connection with liberalism à la Mill or Hayek. It is concerned with the relation between individual preferences and the SWF. There is, undeniably, a conflict between condition L and the Pareto principle, because they both deal in outcomes and set up contradictory criteria for a 'socially better' state of affairs. But this merely reinforces the point that condition L is not a liberal principle.

I fear that all this may appear like a minor skirmish on the borders of social choice theory, the details of which can be of no possible interest to anyone except Sen and myself. I want to emphasize that this is not so. If I am correct, the implication is that we cannot expect to get any help from social choice theory in analysing rights because the whole concept of social choice or social preference is too simple to accommodate the concept of a right. Sen's proffered translation of Mill and Hayek into the language of SWFs comes up with altogether the wrong kind of thing.

Since the world is full of social choice theorists and many of them are very clever, I had better be more specific. I do not see how one could deny in advance that it may be possible to carry out some complicated translation of what we actually want to say into some extension of the language of social choice. What I do feel fairly safe in denying is that the result will illuminate any significant problem in moral or political philosophy. Rather, it will continue to generate its own internal puzzles out of its own inadequacies.

The reason for this is that social choice theory is a theory producing orderings of outcomes (states of the world), whereas the theories we require must include concepts such as the right of an actor (which might be an individual or a collectivity like a governmental subunit or a country) to make certain choices and the notion that an actor ought or ought not to do certain things within the range of things it has a right to do. These simply cannot be connected directly to a system for generating rankings of outcomes, which is what social choice theory is.

Thus, Sen's idea that the solution lies in changing the utility information (so that the lewd and the prude might choose not to have their nosy preferences counted in the SWF) is fundamentally misguided. For the question of the socially preferred outcome (according to the SWF) is different in nature from the question of the rights people should have. There is no need to mess about with the SWF. It may really be that the world would be a better place if the prude widened his horizons and the lewd didn't skim novels for the dirty bits, and they are quite right in agreeing this is better than the reverse. But it does not follow that it is morally permissible for them to bring it about if that can be done only by interfering with one another's freedom. The whole idea of anti-paternalism is, after all, that some consequences admitted to be superior should not be brought about if the means are a reduction of freedom of choice. I believe myself that anti-paternalism has got completely out of hand when it issues in objections to the mandatory use of seatbelts or crash helmets. But hardly anybody would wish to say that a reduction in freedom of choice is *always* permissible (still less required) whenever it brings about better outcomes, and that is all we require for the present case.

The second claim mentioned at the beginning of this section is more substantively interesting, and is one that Sen himself has not explicitly defended, though, as I have noted, he has said that he thinks it can be defended. This is the claim that, when rights have been assigned in an ethically acceptable way, it may be wrong for one person to offer to use his rights in a certain way with the object of inducing another to use his rights in a way he would not otherwise choose. I have suggested that such cases can arise, but they all seem to be subsumable under the objection that the freedom to choose is more apparent than real.

It should be emphasized that a positive response to the second claim in no sense rehabilitates the 'liberal paradox'. Suppose we conclude for some reason that I ought not to offer you $10 never to wear that tie again even though I'm clear that it would be worth it to me and even though you'd be happy to close with the offer. We should not rush to the conclusion that Pareto optimality fails *because* it conflicts with the liberal principle establishing (among other things) your right to wear whatever tie you like. All that can be deduced is that not every Pareto-optimal outcome available under the system of designated rights should be brought about. But the rejection of the view that all Pareto-optimal moves are desirable is a position with respect to the use people ought to make of their rights. It is logically quite independent of the case for establishing rights. We could equally well support exactly the same system of rights and at the same time maintain that the more Pareto-optimal deals are consummated the better.

If we support a set of rights but believe that not all Pareto-optimal moves are desirable, we will be committed to saying that sometimes people do things they have a right to do but that they shouldn't do. If we support a set of rights and believe that Pareto optimality is one criterion of a desirable outcome, we will find ourselves saying that sometimes it would be better if people were to make some Pareto-optimal deal but that they have a right not to if they choose not to. Both positions are consistent.

Even if there is no logical connection there, it may be asked how my rather indulgent views of trades conforming to the Pareto principle can be reconciled with my earlier-expressed view that judgments about better or worse states of affairs (e.g. that it would be better if you didn't read *Lady Chatterley*) should not be treated as preferences and cranked into a SWF. My answer is that there is no inconsistency here because the second question is not whether the fact that the prude thinks it is socially better for the lewd not to read *Lady Chatterley* provides a reason for somebody else (the lewd or anybody else forming a social welfare judgment) to conclude that it would be better for the lewd not to read it. (The prude must of course think it better, but not that it's better *because* he thinks it's better; rather that he thinks it's better because it *is* better.) The second question that I took up is, instead: given that (on the basis of the contingent offers) the prude wants to read *Lady Chatterley* and the lewd doesn't, is there any reason for us (or them) to conclude that they ought not to do so? In other words, this is a case where, *ex hypothesi*, each does have a preference for what the other does, and these preferences are so strong that each cares more about what the other reads than he does about what he reads himself. Given this highly unusual situation, the question is whether there is anything wrong in their doing what they both prefer, so that the prude reads and the lewd doesn't. My answer that this is probably all right, given the preferences, does not conflict with my views that the assignment of rights should be done on quite separate criteria from those on which we judge the use people make of them, and that in general judgments of social welfare should be distinguished from preferences. The lewd and the prude *do* have strong preferences as regards one another's behaviour; but that is no reason for concluding that beliefs about what is better or worse are usually no more than weaker preferences (or, as Hume might have said, calmer passions).

REFERENCES

Arrow, K. (1951) *Social Choice and Individual Values*, New York: John Wiley and Sons.

Barry, B. (1965) *Political Argument*, London: Routledge and Kegan Paul.

Barry, B. (1973) 'Wollheim's paradox: comment', *Political Theory* 1, 317–22.

Coleman, J. S. (1973) *The Mathematics of Collective Action*, London: Heinemann Educational Books.

Godwin, W. (1971) *Enquiry Concerning Political Justice with Selections from Godwin's Other Writings*, abridged and edited by K. C. Carter, Oxford: Clarendon Press.

Grant, G. (1981) 'The character of education and the education of character', *Daedalus* 110, 135–49.

Greene, G. (1980) *Doctor Fischer of Geneva or the Bomb Party*, New York: Simon and Schuster.

Little, I. M. D. (1952) 'Social choice and individual values', *Journal of Political Economy* 60, 422–32. Reprinted in E. S. Phelps (1973) *Economic Justice*, Harmondsworth: Penguin, pp. 137–52.

Mill, J. S. (1859) 'On liberty', in J. M. Robson (ed.) (1977) *Essays on Politics and Society*, Toronto: University of Toronto Press, Vol. 1.

Sen, A. K. (1970a) 'The impossibility of a Paretian liberal', *Journal of Political Economy* 78, 152–7.

Sen, A. K. (1970b) *Collective Choice and Social Welfare*, San Francisco: Holden-Day, pp. 78–88.

Sen, A. K. (1976) 'Liberty, unanimity and rights', *Economica* 43, 217–45.

Sen, A. K. (1979a) 'Personal utilities and public judgments: or what's wrong with welfare economics', *Economic Journal* 89, 537–58.

Sen, A. K. (1979b) 'Utilitarianism and welfarism', *Journal of Philosophy* 76, 463–89.

Waldron, J. (1981) 'A right to do wrong', *Ethics* 92, 21–39.

Wollheim, R. (1962) 'A paradox in the theory of democracy', in P. Laslett and W. G. Runciman (eds.), *Philosophy, Politics and Society*, 2nd series, Oxford: Basil Blackwell, pp. 71–87.

2. The purpose and significance of social choice theory: some general remarks and an application to the 'Lady Chatterley problem'*

AANUND HYLLAND

I. Introduction

On one level, social choice theory is a mathematical discipline; conditions are formulated precisely in the form of axioms, and theorems are proved. But if the theory is to be of any use, the story cannot end there. The results must tell us something about 'real' issues. There are, however, several possibilities as to what these issues might be. Perhaps the theory can be used to evaluate systems for decision-making; perhaps it can give insight into the inherent properties of concepts like 'collective decisions' or 'social welfare'; or perhaps, in a more practical spirit, it can aid us in designing political institutions and constructing actual decision procedures.

My impression is that in many recent works in social choice theory, the purpose of the theory and the significance of the results are not discussed thoroughly enough. Thus the authors run the risk of producing theorems with not much more than mathematical content.[1] This has not always been the case. Kenneth Arrow's book *Social Choice and Individual Values*, from which the whole field can be said to originate, contains an extensive discussion of interpretations and significance, and in the years after its publication, there was consider-

---

* Most of the substance of this paper was presented at the Ustaoset conference on the Foundations of Social Choice Theory, as comments to Brian Barry's contribution. I have found it convenient to write it out as a separate paper, although the last part still contains remarks directly related to Barry's discussion. The paper was written while I was a visitor at the Department of Economics, Stanford University.

[1] I do not claim that this holds for *all* recent works. Nevertheless, I make a rather sweeping statement about the literature, and it would not be unfair to ask that I provide detailed documentation. I shall not do so here, but I can refer the reader to Kelly (1978), who gives a systematic account of recent results in the field, many of which can serve to illustrate what I have in mind. Of course, a paper need not be worthless even if issues of interpretation are not discussed, since its theorems can be of use to others.

able debate on these issues.[2] Later, a large number of purely technical papers have been forthcoming. A charitable explanation for this change would be that all issues of interpretation have already been fully explored, and a general consensus has been achieved concerning the significance of results in social choice theory. Then subsequent authors could build on the consensus without repeating the discussion. But it is quite clear that no such general consensus exists.

Moreover, there is no reason why the theory should be given the same interpretation in all connections; there may be advantages in having several interpretations available. This is analogous to the situation in mathematics, where many modern axiomatic theories derive their strength and usefulness from the fact that the basic concepts, and therefore the theorems, can be interpreted in a variety of ways. (For example, this holds for the theories of groups, rings, vector spaces, topological spaces, etc.) When a theorem has been proved, many results can be derived by applying the theorem to various interpretations of the theory. Thus time and effort are saved. Perhaps more importantly, going back and forth between different interpretations of an abstract theory can result in ideas that would otherwise not have emerged. Social choice theory also has an axiomatic structure; therefore, similar advantages can be obtained by interpreting this theory in more than one way.[3]

The situation is not, however, quite analogous to axiomatic theories in mathematics. To determine whether a given structure is, for example, a vector space, is generally a straightforward matter. Even if it occasionally may be *difficult*, it certainly does not involve issues of judgment. When a theorem of social choice theory is transferred from one interpretation of the theory to another, both the result itself and the conditions (axioms) change their meaning. It is possible that a condition which is reasonable or even compelling in one interpretation becomes less so, or totally meaningless, in another. This underlines the need for being explicit about the interpretation (or interpretations) one has in mind when a formal result is presented and discussed; if this

---

[2] In the preface to the second edition of his book, Arrow writes this about the literature that emerged between 1951 and 1963: 'Some of the new literature has dealt with the technical, mathematical aspects, *more with the interpretive*' (my italic).

[3] This point has already been made by Arrow (1963, p. 87): 'One of the great advantages of abstract postulational methods is the fact that the same system may be given several different interpretations, permitting a considerable saving of time.'

is not done, the result has no other significance than the purely mathematical. And, as already pointed out, a general reference to traditions and customs in the field will not suffice.[4]

In Section II, I present one view of how results in social choice theory can be interpreted. In other words, I give one possible answer to the question of what is the purpose and significance of the theory. As should be clear from the discussion above, I do not consider this the only possible interpretation, but I hope it is a meaningful one. Neither do I claim that the ideas presented are original; rather, I try to put together, in a systematic manner, well-known thoughts and ideas.

Section III contains comments on a specific problem, namely the so-called liberal (or libertarian) paradox, originally presented by Sen and discussed extensively in Barry's contribution to this volume.[5] In line with my previous remarks, I maintain that a meaningful discussion of the 'paradox' is only possible when one has decided how to interpret the basic concepts of the theory within which it is formulated. My discussion will be based mainly on the interpretation presented in Section II, but at the very end I briefly consider another possibility.

II. Evaluating social systems

All the time, we evaluate or pass ethical judgment on social systems and arrangements. We characterize societies as democratic or undemocratic (or more or less democratic), or as 'good' or 'bad' in other respects. Social choice theory can be viewed as an instrument aiding us in making such evaluations. Results from the theory can help us clarify the issues. For example, they tell us which sets of requirements can possibly be met, and which are logically inconsistent. In this section, I shall explain this idea in more detail, thereby giving one view on the significance of the theory.

I said that 'we' pass ethical judgments on social systems. This is not precise enough. An ethical judgment must be *somebody's* judgment.

---

[4] Sen (1977, p. 82) makes essentially the same point. In a footnote, he quotes Arrow (see footnote 3 above) and adds: 'But *that* probably is *also* one of the great disadvantages of these methods' (author's italic). If this is to be taken literally, I disagree; the possibility of varying the interpretation cannot really be a disadvantage. But it is dangerous, since it admits the possibility of confusion. (This is probably also what Sen had in mind.)

[5] See Sen (1970), Gibbard (1974), Sen (1976), and Barry (this volume). Further references are given in the last two of these papers.

For the time being, therefore, we concentrate attention on the judgments of one person, who can be thought of as an ethical observer. To
be specific, let us assume that I am this observer. Thus *I* want to
evaluate certain social arrangements, and social choice theory will be
used to aid *me* in doing so. Later, we shall switch to a less individualistic mode and see if the theory can be of more general use.

The society in question can be a society in the usual sense of the
word, namely all the inhabitants of a country or a subdivision thereof
(or, for that matter, the whole world), together with all the formal and
informal institutions that bind them together. But it can also be a group
that is smaller or has a more specialized purpose, such as an association
or organization. Conceptually, it makes little difference what kind of
'society' we are talking about. (This may, however, be important at a
later stage, when we reach the question whether the ethical judgment
can serve as a guide to action.)

An ethical observer might be interested in many aspects of a society.
The one we shall consider here is, of course, the one studied by social
choice theory, namely the connection between individual preferences
and social outcomes. This can only make sense if we assume that such a
connection exists. Consider the following thought experiment. The
society, with all its formal and informal institutions, is given and fixed.
Take a profile of preference orderings, one ordering for each
individual, and suppose that the individuals actually have these preferences. What will then happen in society? Certainly, *something* will
happen. People will act, presumably in a way that bears some relation
to their preferences, interaction will take place through the social
institutions, and all of this will result in an outcome. We associate this
outcome with the given preference profile. By performing this thought
experiment for every possible preference profile, we establish a
function from profiles to outcomes. This is called the *social decision
function* of the given society.[6] It is a formal representation of how
society functions, and this functioning is the subject-matter for the
ethical judgment.

Below, the definition of social decision function will be modified in

[6] This and related concepts are known in the literature by many names, for example,
social welfare function, constitution, collective choice rule, and social decision function.
Some authors use several of these terms, giving them separate technical definitions. I
have no such specific definition in mind; the term social decision function is chosen
because I feel it best fits the rest of my description.

two respects. But it is convenient to use this formulation for the present, introducing the modifications as we comment on the definition. Quite a few such comments are in order, some of them conceptual and some more technical.

In the thought experiment that defines the social decision function, everything but individual preferences is kept constant. This includes the social institutions that translate individual actions into outcomes. But it also includes the rules according to which individuals' preferences determine their actions. Since people do not know each other's preferences, and perhaps do not completely understand how the social institutions work, they in effect participate in a game of incomplete information. In game theory, alternative assumptions have been proposed concerning people's behaviour in such a situation. These behaviouristic assumptions are to be kept fixed as preferences vary. (To be precise, they shall be kept constant for any one individual; the society we study could be one in which different groups of people choose their actions according to fundamentally different principles.) On psychological grounds, one can question the realism of varying preferences while all these other things are held constant, but I shall not go into a discussion of that issue.

Even under all these everything-else-is-given-and-fixed conditions, it is not clear that every preference profile determines a unique outcome. For example, our behaviouristic assumptions could be based on one of the equilibrium concepts from game theory; that is, we assume that people choose actions that form an equilibrium of a certain kind.[7] For some preference profiles, there may be more than one equilibrium, and these can lead to different outcomes. In such a case, the value of the social decision function shall not be one outcome, but the *set* of all outcomes that *can* come about when preferences are given by the profile that is being considered. This is called the *choice set* for the profile. Thus we have made one modification in the definition. For formal consistency, we must insist that the value of the social decision function is always a set of outcomes. The set has one element if and only if the corresponding preference profile, through the social institutions, etc., determines a unique outcome. It is always non-

[7] This must be supplemented by assumptions about what happens if no equilibrium exists. Since the reference to game theory and equilibria is only used for illustration, I do not go into details on this issue.

empty, corresponding to the fact that some outcome will of necessity result from any preference configuration.

The definition given above in no way implies that the social outcome depends solely on individual preferences. Many other factors can affect the outcome, but we keep these fixed and study only the connection between preferences and outcomes. In fact, it is consistent with the formalism to assume that the other factors alone determine the outcome while preferences play no role; then the social decision function is constant.

For a society of any complexity, it is clearly impossible in practice to find and describe the complete social decision function. In my role as an ethical observer, I shall need to know something about the social decision function, but complete knowledge will not, in general, be required. Still, it seems reasonable to assume that the whole function exists, in an ideal sense; and there are advantages in talking about it as an entity, not just about the pieces of it that are known or need to be known.

Should one be concerned only with the connection between the one profile of individual preferences that actually exists and the resulting outcome, or with the whole (hypothetical) function linking preferences to outcomes? There has been a lot of discussion of this issue.[8] The fact that I talk about such a thing as a social decision function could indicate that I support the latter point of view. This I indeed do, but I do not believe that adherents of the former viewpoint can avoid explicit or implicit consideration of a function defined on the set of preference profiles. It is possible to discuss one specific society, concentrating on facts and avoiding all mention of general principles. But as soon as one moves on to general issues, as supporters of both points of view certainly do, the discussion must take place in a setting where individual preferences are not specified. Hence different possible preference profiles, and the outcome resulting from each of them, must necessarily be considered; that is, a functional relationship is being studied.[9] This is not to say that the answer to the question asked at the beginning of the paragraph is unimportant. When it comes to imposing conditions on the social decision function, it makes a dif-

[8] Arrow (1963, pp. 104–5) comments on some of the earlier discussion.

[9] A remark by a proponent of the single-profile school is illuminating. Samuelson (1967, p. 49), in discussing Bergson's (and his own) approach, says this about the preference profile: 'It could be *any* one, but it is *only* one' (author's italic).

ference whether one considers only the connection between actual preferences and actual outcome in a society, or the whole functional relationship between preferences and outcomes. What are often called interprofile conditions, that is, conditions saying something about how the outcome (or the choice set) shall change when preferences change, make no sense in the former case, but are meaningful in the latter.[10] (The discussion in Section III involves no interprofile condition, so the distinction makes no difference there.)

Functions studied in social choice theory often have as their values preference orderings (interpreted as society's preferences), rather than outcomes or sets of outcomes. The difference is largely a technical one. This is certainly true if we make another modification in the definition of a social decision function. In the thought experiment that determines the function, we vary not only individual preferences, but also the set of outcomes that are feasible, this being a subset of the set of conceivable or theoretically possible outcomes. Thus the argument of the function consists of two parts, a preferences profile and a set of outcomes, called the feasible set. The corresponding value of the function is the choice set, that is, the set of outcomes that can result when individual preferences are the ones described by the given profile and the feasible outcomes are exactly those in the given set. Of course,

[10] Many well-known results in social choice theory, such as Arrow's impossibility theorem, can be 'translated' into a single-profile setting. That is, a formally similar theorem can be proved using only one profile. See, for example, Parks (1976), Kemp and Ng (1976), and Roberts (1980); Roberts (1980) also gives a review of results of this type. Some of the papers published on this subject can be used to illustrate the danger of translating a result from one interpretation of the theory to another. (See footnote 4 above and accompanying text.) Compare, for example, Arrow's theorem to that of Parks (1976). In Arrow's model, a dictator is a person whose preferences prevail no matter what other people's preferences are. Then it is not difficult to agree that the existence of a dictator is undesirable. In the single-profile model, a dictator is a person whose individual preferences coincide with the social ones in the one and only profile under consideration. Nothing is necessarily wrong with that; the decision process can be perfectly democratic, and one person simply turns out to be on the winning side on all issues. (This point was made by Jon Elster.) In the paper cited, other assumptions allow us to rule out this possibility, but this is not pointed out by the author, who seems not to appreciate the fundamental difference between the dictatorship concepts of the two types of model. The condition of unrestricted domain in Arrow's sense can, for example, be given a normative justification; a decision procedure *ought to* work no matter what people's preferences are. (Other possibilities exist. In fact, I argue differently in the main body of this paper; see Section III.) The corresponding condition in a single-profile model can only be interpreted as a statement about the preference profile that is being studied; that is, an assumption is made about the preferences people actually have. In my opinion, the condition, thus interpreted, is not reasonable. The point, however, is not my view, but the author's failure to discuss the issue.

the choice set must be a subset of the feasible set. Now we can consider the behaviour of the function, when the feasible set varies while the preference profile is held constant. If this behaviour satisfies certain rationality conditions, it is possible to define an equivalent function whose argument is a profile of individual preferences and whose value is a social preference ordering.[11] In this sense, there is only a technical difference between having, on the one hand, preference orderings, and, on the other, sets of outcomes, as values of the function. (Readers who are confused by this paragraph can ignore it. The issue discussed here is of little importance for the rest of this section, and totally without importance for Section III. I raised it only to connect my approach to the tradition within the theory.)

Outcomes and preferences have figured prominently in the discussion, but so far little has been said about the nature of these entities. An outcome is a complete description of everything that happens in society, or everything that is relevant to the ethical observer's judgment. When the word society is taken in its usual meaning, an outcome is an immensely complicated thing, and the set of possible outcomes has an enormous number of members. This reinforces the earlier observation that we cannot expect to describe the social decision function in complete detail. If the 'society' is a smaller special-purpose group, the factors relevant for the ethical judgment can perhaps be limited in such a way that the description of one outcome and of the set of all outcomes becomes manageable.

Individual preferences are, of course, preferences over the set of outcomes. No specific assumption is made about the form of the preferences. They can be purely ordinal (an individual simply ranks the outcomes from best to worst, possibly with ties) or cardinal (a person is, in some way, able to say how much better one outcome is than another). Interpersonal comparison of preferences may or may not be possible, that is, statements of the type 'outcome x is better for person i than y is for j' may or may not be considered meaningful. The traditional assumption in social choice theory is that preferences are ordinal and non-comparable, but my general discussion applies equally well to the other possibilities mentioned. (Moreover, the distinctions make no difference for the problem considered in Section III.)

[11] For details, see, for example, Blair et al. (1976).

Nothing is said about motivation behind preferences. To say that outcome x ranks above y in a person's preferences simply means that this person, all things considered, would rather see x happen than y. Preferences can be based on egoistic concentration on private consumption, on altruism, on ideal principles of ethics, or on any combination of these factors. We do not distinguish between these possibilities. Critics might object that this amounts to throwing away relevant information. They can argue that what happens in a society ought to depend not only on people's preferences, but also on why they have these preferences. Since the whole exercise is concerned with ethical judgments, this is a relevant objection. My response is that the definitions do not imply that only preferences determine outcomes; people's motivations can be included among those 'other things' that are relevant, but are kept constant in the definition of a social decision function. The critics can question the realism of varying preferences while holding motivation constant. Moreover, they can claim that my response is an evasion rather than a defence: given my assumptions, the ethical observer will frequently be unable to say whether a certain social decision function is 'good' or 'bad', since the judgment depends on the unspecified motivational factors. I stop the hypothetical dialogue here. The inability to distinguish between equal preferences differently motivated is an inherent weakness of social choice theory, not just of my interpretation. Whether it is an important weakness, I shall not discuss here.

By letting preferences range over outcomes, we implicitly assume that individuals are concerned only with 'the business end of politics'. In practice, people have opinions and feelings not only about the outcome, but also about the process through which the outcome is determined. If we identify the process with the social decision function, this amounts to saying that the function should be included as an argument to the preference relations. A person may prefer outcome x to outcome y in any given society, but still prefer living in a society characterized by social decision function f and getting outcome y, to the combination of function g and outcome x. (Example: I prefer losing in a democratic decision process to being the dictator.) This cannot be captured by the model. A suggestion for solving the problem is to incorporate the decision process into the specification of an outcome. But the difficulty re-emerges one step removed, since people can have preferences for the process by which the process is chosen. The definition of social decision function makes it impossible to 'close the loop' and include processes of all orders in the outcomes, since the set of outcomes must be specified before we can talk about preferences, and hence before the function can be defined. The severity of the problem depends on the extent to which people actually have feelings and opinions about the process. Perhaps such feelings are less common and weaker when it comes to processes for choosing processes, or entities of even higher order. If this is so, the problem can at least be alleviated by making the first-order process, or processes of the first few orders, a part of the outcome.

This ends the comments on the definition of a social decision function. To recapitulate, I place myself in the role of an ethical observer, wanting to evaluate a society. The society is characterized by a social decision function, which may be known only partially.

I formulate a series of properties that I ideally want society to possess. These represent aspects of 'democracy', 'justice', 'efficiency', or other generally acclaimed but vague concepts. If I observe that any of these properties is not satisfied, I provisionally form the judgment that there is something wrong with the society.

Each property can be translated into a condition on the social decision function. It then becomes a statement in the language of social choice theory. I can use the whole body of results from the theory to deduce the consequences of my list of conditions. It is quite possible that I regard some of these consequences as undesirable, or that the conditions turn out to be logically inconsistent. In the second case, I have to reconsider the original list of properties, to see which of the items are least compelling and can most easily be omitted or weakened. In the first case, I also want to see if anything should be modified. The modifications lead to a new list of properties, the consequences of which can be deduced. I repeat the whole process, and I hope to end up with a set of properties that is consistent and free from undesirable consequences. (Or, more realistically, the consequences, although perhaps undesirable on the surface, are in my considered judgment lesser evils than giving up any of the properties.)

Inputs to the process described above are the properties I originally formulated. These can be considered prima facie ethical judgments. There is also another set of inputs, namely the intuition that allows me to characterize certain consequences as undesirable, that is, intuition about particular aspects of a social decision function. No one set of inputs has definite priority over the other. I use both, together with results from social choice theory, to come up with an informed judgment, represented by the consistent list of properties that emerges in the end.[12] It should now also be clear how social choice theory can serve as an instrument aiding me in making ethical evaluations.

Thus I have reached final judgment on the given society: if the conditions in the final list are not satisfied, things are not as they ought to be. Typically, this judgment will be less harsh than the provisional one. My standards have mellowed because they were confronted with the knowledge generated by social choice theory.

The theory is full of impossibility theorems, that is, results to the effect that a certain set of conditions cannot all be satisfied. Usually, such a theorem is thought of as something negative, something that creates a problem. But there is also a positive side to the story, which is stressed by my interpretation: Before the theorem is proved, we look around and conclude that no existing society, nor even any proposed social arrangement, satisfies all the conditions we see good reasons for imposing. This, we might think, must be due to human intellectual or moral fallibility. Then the impossibility theorem is proved. We conclude that there is, after all, no reason to worry about the observed state of affairs. Things have to be this way because of the laws of logic, and our observations do not prove that anybody is foolish or corrupt.

The final set of conditions must be consistent. This imposes, in a sense, an upper limit on the strength of the conditions. There is no similar lower limit. In particular, there is no reason to require that the conditions determine the social decision function uniquely. Lack of uniqueness simply means that different social arrangements are all considered ethically acceptable, and there can be nothing wrong with that.[13] If, at some point during the process, the theory tells me that my present conditions admit more than one social decision function, then I should go back to the properties that determined the conditions, think about them again, and see if anything ought to be strengthened

---

[12] The process I describe here is closely related to the concept of reflective equilibrium discussed by Rawls (1971).

[13] In other interpretations of the theory, for example, when it is used as a tool for constructing decision procedures, uniqueness might possibly be an issue.

or added. But at the final stage this has already been done, and there is no reason to insist on uniqueness.

As I have described it, the ethical judgment simply takes the form of approval or disapproval of a social decision function. This may be regarded as a rather meagre output of a long and complicated process. Ethical judgments ought to have more nuances than just 'good' and 'bad', and when it is concluded that something is wrong somewhere, the obvious next question is: what should be done about it?

The first objection can easily be answered. Nothing prevents me from applying several standards of varying strictness. Each standard corresponds to a set of conditions, and each must be subjected to the tests and possible modifications described above. Results from the theory can be used to check whether one set of conditions is really less strict than (that is, implied by) another. Armed with such a series of sets of conditions, I can make judgments on a finer scale than described above.[14]

The second issue is more complicated. The analysis can generate ideas for constitutional or institutional reforms in the society under consideration. Since the social decision function depends on informal as well as on formal institutional arrangements, it need not be immediately clear whether and how such reforms can be implemented. In other words, the analysis tells us that decision-making in the society has certain undesirable properties that can logically be avoided, but it does not tell us how they can be done away with in practice. The severity of this problem depends on what kind of 'society' we have in mind. If it is a well-structured organization with a specific purpose, the difficulties need not be great; reforms can be achieved by amending the by-laws. For a society in the usual sense of the word, things are usually more complicated.

In my opinion, ethical judgments, as described here, have value in themselves. Of course, being able to correct what has been found wrong would be even more valuable. There is no contradiction

---

[14] For example, all conditions but one can be kept constant, while the remaining one is successively weakened from one set to another. Alternatively, one could try to construct a continuous measure of the degree to which a social decision function satisfies a certain condition (or a set of conditions). Usually, it is not easy to do this in a precise and meaningful way. Therefore, this approach has not often been adopted in social choice theory, but it might be suggested as an area for future research.

between my approach and this more demanding one. (Similarly, the same formal theorems can be used in my interpretation, and in one where social choice theory is used as an instrument for constructing decision procedures.) But in this paper I shall not attempt to discuss the array of difficult issues related to implementation of reforms.

Social choice theory does not *make* ethical judgments. The judgments are *mine*, in my capacity as an ethical observer. All inputs to the process come from me; they are either general principles or properties that I find compelling, or my intuition about particular cases. Reconsideration and reconciliation of the original conditions must also be based on my own ethical intuition, perhaps intuition on a higher level. The only thing the theory does, is to clarify the issues by pointing out consequences and inconsistencies. In particular, it tells me where the line between the possible and the impossible is.

Is it really necessary to go through the whole process of defining a social decision function and testing for general properties? Cannot ethical intuition be applied directly, without invoking an extensive theoretical apparatus?

One answer to this has already been given: direct use of intuition can easily lead to the imposition of conditions that cannot be met, or can be met only at an excessive cost in other respects. But there is another line of response, related to the question whether the ethical judgment has to be purely individual, as assumed so far, or whether a higher degree of universality is possible. If each member of a group serves as an ethical observer, I believe it is easier to obtain agreement on a set of general conditions than on direct evaluations. In particular, this holds if the observers are also members of the society being observed. When one is forming a judgment directly, it is almost impossible to disregard one's own particular interests and actually evaluate the social arrangements. General principles are necessarily farther removed from each person's specific situation. When going through the process described above, the individuals know their own preferences and positions in society. But these can change in unknown ways, while ethically determined conditions on the social decision function are supposed to be permanent. Hence discussion of such conditions takes place in an environment where people are not fully aware of how various alternative conditions will affect them. This should make agreement more likely. It is assumed here that each person in principle goes through the whole process described above, but there is public discussion of properties and conditions, and people are free to try to influence each other's judgments. I do not claim that agreement is certain under these circumstances, only that it is more likely than if judgments are formed directly and intuitively.[15]

The devil's advocate will ask why agreement is desirable. To answer this, I have to move outside my own model. If we are to make decisions about reforming the society whose social decision function has been studied (or, more generally, make decisions about decision-making), we easily run into the same difficulties that motivated the study of decision-making in the first place. But these difficulties evaporate if there is substantial agreement about the properties a social decision function ought to possess, and hence there are clear advantages to achieving agreement. Even if I stick to the original formulation, in which my judgment as an ethical observer is the only issue, the arguments of the previous paragraph have relevance. If I am a member of the society being studied, I will have difficulties in distinguishing my preferences, that is, my views on what ought to happen, from my evaluation of social institutions and arrangements. Therefore, although there is nobody with whom I need to agree, consideration of general principles, and thus the use of social choice theory, is not just an unnecessary complication.[16]

Questions can be asked and objections raised on many points above. I have raised some of these issues, but in order that the discussion should not get completely out of hand, much has been left out.[17] Still, I hope to have demonstrated that social choice theory has

---

[15] This argument resembles the 'veil of ignorance' construction of Rawls (1971). But Rawls carries the idea much further than I do; he postulates a situation in which people know nothing about their particular positions in society, and he claims that complete agreement on moral principles is then inevitable. Thus we can say that I employ a *veil* and Rawls an impenetrable *wall* of ignorance.

[16] I shall mention, but not discuss, a possible objection to the last two paragraphs: Ethical intuition is developed through experience. When discussing general principles, we move far away from the type of situation on which intuition is based; therefore, our judgments become less reliable. Hence statements about general principles and conditions should be given less weight than statements about specific, familiar cases.

[17] I intend to discuss some of the objections more thoroughly elsewhere; see Hylland (forthcoming).

substantive content, by identifying a class of interesting problems to whose solution the theory can contribute.

III. The liberal paradox

The liberal paradox is a theorem of social choice theory; certain conditions are shown to be logically inconsistent. The theorem was proved by Amartya Sen. In the interpretation of the theory given in Section II, the consequence is that a social decision function cannot be required to satisfy all these conditions; at least one must be given up or weakened. In this section, I shall discuss the conditions and ask which of them can most easily be dispensed with. The more difficult it is to answer this question, the more important and interesting is the theorem. On the one extreme, if I conclude that at least one of the conditions is not compelling at all and should not be imposed even if there were no problems of consistency, then the theorem is without interest. (Still it may be important in other interpretations of the theory.) On the other hand, if I am unwilling to give up anything and insist, even after struggling with the issues for a long time, that the conditions are all basic and indispensable principles of ethics, then the theorem has brought to light a fundamental contradiction in my thinking and is therefore of great interest.

The theorem is discussed in detail in Section I of Barry's contribution to this volume, to which the reader is referred. The conditions shown to be inconsistent are unrestricted domain, Pareto optimality, and liberalism, plus a requirement of collective rationality. The conditions must be rephrased slightly, since Barry discusses functions whose values are social preference orderings, while the values of a social decision function in my sense are choice sets. This change makes no substantive difference, however.

Unrestricted domain means that the social decision function must be defined on all profiles of individual preference orderings; this is implicit in the definition given in Section II. Collective rationality requires that the choice set be non-empty, which must necessarily hold in the present interpretation.

For a given profile of preference orderings, we say that an outcome x *Pareto dominates* an outcome y if every individual prefers x to y. An outcome is *Pareto-optimal* if no feasible outcome Pareto dominates it. The Pareto optimality condition is this: for every preference profile, the choice set shall consist solely of Pareto-optimal outcomes. (In technical terms, the choice set shall be a subset of the set of Pareto-optimal outcomes.) Equivalently, if x is feasible and Pareto dominates y, then y shall not be in the choice set. Nothing is said about x.[18]

For an individual i and outcomes x and y, the social decision function is said to *respect i's right over x and y* if the following holds: for every preference profile in which i prefers x for y, if x is feasible, then y is not in the choice set; and for every profile in which i prefers y to x, if y is feasible, then x is not in the choice set. The liberalism condition requires that the social decision function respect the rights of at least two persons over at least one pair of outcomes each.[19]

Given the pairs of outcomes over which rights are respected, it is now easy to construct a preference profile such that every outcome is excluded from the choice set by either Pareto optimality or liberalism. Since this set cannot be empty, we have a contradiction, and the theorem is proved. Several cases must be considered, depending on the extent to which the given pairs of outcomes overlap, but they are all straightforward. (Barry gives the proof for the principal case, where there is no overlap.)

Collective rationality is not, in the interpretation of the theory used here, a normative requirement that can be relinquished if we decide that other conditions must be given higher priority; it is an undeniable fact that something happens in a society no matter what people's preferences are. Unrestricted domain is also a consequence of the definitions, but it is not, in the same way, beyond discussion. If it is possible to conclude, on theoretical or empirical grounds, that certain preference profiles cannot occur, then there is no reason to insist that the social decision function be defined on these profiles. They should be eliminated before we perform the thought experiment that defines

[18] The concepts defined here are *strong* Pareto dominance and *weak* Pareto optimality. An outcome x weakly dominates y if at least one person prefers x to y, while nobody prefers y to x. This gives a stronger concept of optimal outcome (that is, fewer outcomes are optimal) and a stronger optimality condition. Since the weak condition is sufficient to prove the theorem, there is no reason to use the strong one.

[19] According to this definition, respecting somebody's right over x and y is a symmetric property in these two outcomes. An alternative version of the condition, involving only one-sided rights, also suffices to prove the theorem; see Theorems 2–3 and 2–4 in Kelly (1978).

the function, and then unrestricted domain is violated. Note that it is not the unrestricted domain condition itself, but its rejection, which is a substantive assumption requiring justification.[20]

For a given system of rights, the proof of the liberal paradox uses only one preference profile (though it is a different one for different rights systems). To avoid the contradiction, we shall have to eliminate from the domain of the social decision function every profile that can be used to prove the theorem. Sen and others have constructed entirely reasonable examples in which preferences are such that the theorem can be proved. To rule out all these profiles seems out of the question. Pareto optimality and liberalism remain, and the conclusion is that a social decision function has to violate at least one of these. It does not follow, however, that unrestricted domain can now be removed from consideration. I have concluded that this condition cannot reasonably be weakened so much that this alone removes the contradiction; it is still possible that my final solution will involve some domain restrictions, combined with the weakening of other conditions.

Pareto optimality is certainly a compelling principle. If everybody in a society prefers x to y, and y is chosen although x is available, there should be no basis for not switching from y to x. This is essentially what the condition says.

It must be emphasized that imposing the condition does not imply a judgment that every Pareto-optimal outcome is good and every non-optimal one is bad, nor that any Pareto-optimal outcome is preferable to any non-optimal one. The only implication is that a non-optimal outcome cannot be ideal. Sometimes one hears arguments to the effect that once a social institution has been shown to prevent movement from the present situation to one that Pareto dominates it, an adherent of the optimality condition must necessarily support the abolition of the institution. But this reasoning must be based on a misunderstanding of the implications of the condition. To illustrate my point, I shall use an example mentioned by Barry, namely minimum wage laws. Clearly, such a law prevents deals that are advantageous for all those involved, and we can assume that if these deals are struck and everything else is left unchanged, then society moves to an outcome

[20] To reject unrestricted domain, we must be able to deny Samuelson's claim (see footnote 9 above) that the preference profile could be *any* one.

that Pareto dominates the one previously realized.[21] But repealing the law will not leave everything else unchanged. Bargaining power in the labour market will shift, and the final outcome will probably neither dominate nor be dominated by the present situation. Thus the Pareto optimality condition does not determine whether there ought to be a minimum wage law. We have identified a possible improvement from the situation where this law is in effect, but we do not know how to get there; simply repealing the law will not do. (Perhaps it is impossible to implement the change; then we have not found any feasible outcome that dominates the present situation, which may be Pareto-optimal after all.)

The weakness of the condition is further underlined by the fact that *everybody's* preferences have to coincide before it becomes operative. Many examples in which Pareto optimality at first sight appears to have counter-intuitive consequences disappear when this is properly taken account of. Suppose, for example, that persons i and j consider entering into an agreement abolishing i's freedom of speech, in return for some compensation from j. Moreover, assume that i and j, for whatever reasons, prefer that such an agreement take effect. So one might conclude that Pareto optimality requires that the agreement be signed and enforced. Many will find this consequence offensive, and it is natural to think that there must be something wrong with the Pareto condition. But here the effect of the agreement on people who are not party to it has been ignored. They will be deprived of the benefits of i's participation in intellectual and cultural life; therefore, they probably have a preference against the agreement. Hence the Pareto optimality condition implies nothing, and it cannot contradict anybody's intuition. What if i and j are the only members of society? Then Pareto optimality has a definite consequence, which may still be deemed counter-intuitive. If so, I contend that intuition will have to yield. Intuition is developed through observation of actual societies, which have far more than two members. When confronted with an artificial example, such as a two-member society, we have difficulties in adjusting and internalizing the stated assumptions. Therefore, intuitive

---

[21] This assumption is easier to defend if we consider weak Pareto dominance; see footnote 18 above. Even then it can be questioned, since people not involved in the deals can prefer their not being made. If this is true, Pareto optimality has no implications at all in the example.

conclusions are unreliable; they may be influenced by unconscious thoughts of the type 'there really has to be somebody else around, who is hurt by the agreement'.[22]

Social choice theory is usually thought of as a static theory. The members of society are those currently living and belonging to it, and the preferences considered are their present ones. But actions taken today often affect future generations, and people's preferences change over time. Nothing in the formalism prevents us from including among the 'individuals' unborn persons and several time-instances of the same person, each with separate preferences.[23] Expanding the set of individuals in this way weakens the Pareto optimality condition even more. Doctor Fischer's Bomb Party, Section II of Barry's paper, can illustrate this point. Dr Fischer and his guests all prefer that the party be held. If they, with their present preferences, are the only members of society, preventing the party would violate Pareto optimality.[24] But if we include the guests' past and future preferences, things may be different. We are told that Dr Fischer has 'corrupted' the guests, which presumably means that he has deliberately influenced the development of their preferences. Some time in the past they would not have approved of the party. In the broader set of preferences, therefore, there is no unanimity. An ethical observer can, without contradiction, object to the party taking place and endorse Pareto optimality.

The reader may have received the impression that I will defend the Pareto optimality condition under all circumstances (though much of the defence has consisted in showing that the condition implies very little). But this is not really the case. For one thing, I have not said that an ethical observer ought to consider the expanded set of preferences described in the last paragraph, or that I will do so if I am the observer. I have only pointed out that this possibility is consistent with the model

---

[22] See footnote 16 above, where I suggest that one can object to the use of general conditions on essentially the same grounds. In my opinion, the argument is more relevant in the case discussed in the text.

[23] We cannot know future preferences (of living or unborn persons). This is not, however, a valid objection. We do not know current preferences either, and such knowledge is not required in the interpretation of Section II. What must be known is the social decision function, or at least certain aspects of it. (Moreover, we can probably make some fairly safe assumptions about future preferences, but clearly less is known about them than about current ones.)

[24] The same is true if society contains other people not involved in the affair, provided that they are indifferent and strong optimality is imposed; see footnote 18 above.

presented in Section II; it also has its disadvantages, and I am uncertain whether it is a reasonable way of looking at things. Moreover, Barry gives other examples than the bomb party, and I find some of them quite convincing. (Here I refer to examples of the second and third kind; see the last part of Section II in his paper.) Below, in the discussion of liberalism, I shall give another example, in which I am at least willing to consider the possibility of giving up Pareto optimality.

Yet another objection to the condition should be mentioned. It is one thing to say that if everybody prefers x to y then x is, in some sense, socially better than y. But it is something else to claim, on ethical grounds, that there is something wrong with a society in which y occurs. In my interpretation of the theory, I do the latter when I impose the condition (assuming that x is feasible). If y is brought about because social institutions prevent people from cooperating, by placing them in a situation where individually rational actions produce a dominated outcome (as in games of the Prisoners' Dilemma type), this is entirely appropriate; the justification for the condition is exactly the desire to condemn such institutions. Consider, however, the following example: Everybody prefers x to y. Because of the way things work, one person, i, unilaterally controls the choice between x and y, which are both feasible. In full knowledge of this, i acts in such a way that y results.[25] Perhaps we can conclude that i is stupid, but is there any basis for criticizing the social institutions? One response consists in denying the possibility of such an example: the fact that i chooses y shows that preferences are not the ones postulated. Under the interpretation of preferences traditionally used in economics, this is certainly true. But if we assume that i, in addition to preferences for outcomes, also has opinions about procedures and institutions, the example no longer contains a contradiction. (This possibility was discussed in Section II.) More generally, objections can be raised to the identification of preference and choice implicit in much of economic theory, as has been forcefully argued by Amos Tversky.[26] What this shows is that there are inherent difficulties in the preference concept. Then it is not

---

[25] If strong optimality is used, the example can be made somewhat more credible; we can assume that everybody but i is unaffected by, and indifferent to, the choice between x and y.

[26] See quotation in Elster (this volume, pp. 107–8).

surprising that there are difficulties with Pareto optimality. I did not pursue this issue in Section II, and I shall not do so here.

Turning to liberalism, let me first dispose of a secondary issue. Barry argues that the condition, as formulated by Sen and used in the theorem, has nothing to do with liberalism as advocated by John Stuart Mill and other famous liberals. This may be true, but it is irrelevent for my discussion. (At most, it shows that the condition should be given another name.) The real problem is whether the condition expresses a compelling principle of ethics, not why it is compelling.

There are many cases in which intuition strongly suggests that one person ought to have complete control over certain choices. The question whether this supports the imposition of the liberalism condition must be discussed with the model of Section II in mind. Suppose we have concluded, on intuitive grounds, that person i ought to determine the choice between x and y. Assume, moreover, that i prefers x to y, x is feasible, and we observe that y occurs. Does this give me reason for criticizing the social institutions, as I do if I insist that the social decision function respect i's right over x and y (in the technical sense defined above)? Not necessarily, since it is possible that i controlled the choice between x and y but chose y deliberately, in order to fulfil a mutually beneficial agreement with other people, without which the outcome would have been worse for i than both x and y. In general, there is no reason to object to people making such agreements about their right. There are, however, exceptions. In some cases I, and probably most other people, feel that a person should not be allowed to trade away a right. We express this feeling by calling the right inalienable. There is no reason to require that a social decision function should always respect an alienable right; only for the inalienable ones is this a justifiable condition. In this respect, I agree with Barry.

When do we declare a right to be inalienable? One reason could be that being able to trade in the right weakens the holders' bargaining position and hence is disadvantageous to them. (In the absence of such indirect effects, an alienable right is always more, or at least not less, valuable than an inalienable one; the former gives everything the latter gives, plus possibly the opportunity to enter into advantageous agreements.) Another possible reason is that one person's trading away the right can be harmful to others. In these two cases, requiring that the right be respected does not contradict Pareto optimality; see the discussion above of examples involving minimum wage laws and freedom of speech.

To me, it seems clear that the right to read, or not read, one specific book should not be considered inalienable. This contrasts, perhaps, with the general right to read books of one's choice, which is more like freedom of speech. Similarly, if I were able to sleep on my back rather than on my belly, and somebody offered me a satisfactory compensation for doing so, I can see no reason why other members of society or an ethical observer should object (except possibly on the grounds that such an offer must be considered evidence of insanity on the part of the compensator, combined with a rule that I am not allowed to take advantage of the insane). In other words, to find examples of inalienable rights that conflict with Pareto optimality, we need stronger meat than *Lady Chatterley's Lover*.

Gibbard (1974, p. 398) offers an example in which the issue is the right to marry or remain single. This is clearly less trivial than the cases alluded to in the last paragraph. The example involves three characters, Angelina, Edwin, and the judge, though the latter plays a rather passive role. Gibbard describes the preferences this way: 'Angelina wants to marry Edwin but will settle for the judge, who wants whatever she wants. Edwin wants to remain single, but would rather wed Angelina than see her wed the judge.' Edwin's first preference is to remain single, but if he insists on his right to do so (a right which is not questioned by anybody, of course), the end result will be that Angelina and the judge get married. This outcome is not Pareto-optimal, since it is dominated by the one in which Angelina and Edwin get married. (The latter outcome is preferred to the former by both Angelina and Edwin, and hence also by the judge, who wants whatever Angelina wants.) To guarantee Pareto optimality, an agreement between Angelina and Edwin to marry each other is needed; in other words, Edwin must relinquish his right to remain single. Considering the final consequences of accepting or rejecting the deal, he concludes that he wants to accept it. I am not convinced that there is any reason to prohibit such an agreement, although rights of a rather fundamental kind are involved. In discussing Gibbard's paper, Sen (1976, p. 226) suggests a different set of motivations consistent with the preference structure of the example, and he argues that motivation is relevant for ethical evaluation of the outcome. We may deplore the fact that people base their preferences on motivations like those described by Sen. But the issue here is whether I, as an ethical observer, should criticize society for permitting them to strike a deal. I think not.[27] (A condition, of course, is that they act freely and are fully informed of the consequences of the agreement.)

I believe we have the best chance of identifying inalienable rights by considering the ultimate issue, namely questions of life and death. An example will be presented along these lines. I take it for granted that risking one's own life in order to save the lives of a small number of people to whom one has no special obligations is a laudable action, but nobody is morally obliged to do it. Assume that a few people (for definiteness, we suppose there are three of them) are threatened with certain death and can themselves do nothing about it. Two other persons, i and j, are in the situation that either of them can save the ones in danger, but the rescuer risks being killed. Nothing is gained by i and j both participating in the rescue operation. There are three relevant outcomes: the three are saved by i (this we call I); they are saved by j (J); they are not saved (O). Person i thinks that neither i's nor j's life ought to be put at risk, but i is altruistic with regard to j. Hence i's preference ordering is O, I, J. On the other hand, j thinks that the number of lives lost ought to be minimized, but is egoistic in the event of equality according to this criterion. This gives the preference ordering I, J, O. It seems clear that i ought to determine the choice between I and O, and j the choice between J and O.

The example is isomorphic to (that is, it has the same formal structure as) the Lady Chatterley example. In Barry's terminology, the prude corresponds to i and the lewd to j, and reading the book corresponds to rescuing the three persons. The example was deliberately constructed so as to obtain this isomorphism. The preference profile and rights structure can be used to prove the liberal paradox. Moreover, the preferences seem credible and not morally objectionable.

What actually happens in the example depends on the order in which

[27] Sen continues the example by telling us how Edwin can convince himself that he should not accept the deal, in spite of it being advantageous according to his preferences. Then we are into the problem of people having preferences not just for the outcome, but also for the procedure; this issue was briefly considered in Section II.

i and j act and on their ability to cooperate. But the issue here is which outcomes are ethically acceptable. If i and j exercise their rights without cooperation but with knowledge of each other, J will be the outcome. However, J is Pareto dominated. (This assumes that everybody but i and j, including those saved, also prefers I to J, which should not be difficult to accept. If necessary, we can add an assumption that i has a slightly better chance than j of surviving when saving the three.) Should the rights be considered inalienable, in which case no objection can be raised to the non-optimal outcome J being brought about? I am not sure what to say to this, but I am more inclined to answer yes here than in the examples previously considered.

The agreement necessary to move from J to the Pareto dominating I takes the form of i promising to save the three, in return for j's promise not to do so. (To make the example realistic, I shall probably have to tell a story in which i and j must act quickly for the rescue operation to be effective. Hence talking about contracts strains credibility, but I ignore this problem.) Such an agreement is clearly not legally enforceable. After having agreed, i can safely refuse to deliver. The result is O, which is i's most preferred outcome. Foreseeing this, j will not accept the agreement, and we get back to the non-optimal J. By making the agreement unenforceable, the law, in effect, says that the rights involved are inalienable. It is perhaps somewhat beside the point to mention this; I do not discuss which rights *are* inalienable, but which ought to be. Nevertheless, the observation shows that I am not alone in voicing objections to an agreement to move from J to I.

If we disregard the last example, the only inalienable rights that have been identified do not contradict Pareto optimality. Since only inalienable rights must be respected unconditionally by the social decision function, this suggests a way out of the liberal paradox: Pareto optimality should be maintained and liberalism weakened so as not to conflict with optimality, but it should not be totally eliminated.

A solution of this type is presented by Gibbard (1974). He starts with a system of rights, which ideally ought to be respected.[28] If preferences are such that a right conflicts with Pareto optimality and other people's rights, the right is waived. Gibbard's weakened liberalism condition

---

[28] The rights considered by Gibbard in his general model are one-sided (see footnote 19 above), but that makes no essential difference.

requires that the social decision function respect all rights that are not waived; nothing is said about rights that are waived. Thus Gibbard deals with alienable rights, and he also uses that word.

It must be emphasized that the waiving of a right is not an action taken by the person whose right is considered. The actor is the ethical observer, who concludes that although the right generally ought to be respected, there are occasionally good reasons for waiving this requirement on the social decision function.[29] Even if a right is waived, it may be respected. The point is only that its being respected is no longer a part of the (weakened) liberalism condition.

The strength of the revised liberalism condition depends on the initial system of rights. Gibbard considers the following case. There are a number of issues, each of which can be solved in several ways. To each issue, therefore, corresponds a set of possible solutions. An outcome is a vector consisting of one solution to each issue. In technical terms, the set of outcomes is the Cartesian product of the solution sets for the issues. For every person there is an issue which is such that the person has a right over any pair of outcomes that differ only on this issue. (The right is alienable, of course; it can be waived as described above.) Thus we can say that each person controls one issue. In addition, there may exist purely public issues.[30] This rights system is certainly not trivial. (We can, for example, compare it to the original definition of liberalism, where only two persons were required to control one pair of outcomes each.)

Gibbard proves that his liberalism condition, based on the system of rights just described, is consistent with Pareto optimality and the other conditions used in the liberal paradox. This would not be very interesting if rights were always, or almost always, waived. Another theorem shows that this is not the case; waiving occurs rarely and only under special circumstances. Thus having an alienable right in Gib-

---

[29] Kelly (1978, Ch. 9) discusses Gibbard's paper on the assumption that the waiving of a right is an action by the holder. But this interpretation is not really consistent with the rest of Gibbard's model, where preferences are assumed to be known and all problems of preference reporting and incentive compatibility are ignored. Sen (1976, p. 224) considers both a 'pragmatic' and an 'ethical' interpretation of Gibbard; the latter agrees with mine.

[30] Since the set of outcomes is a Cartesian product, one might get the impression that this model rules out technological interdependency between people's exercising of their rights. This is not true, however. The feasible set can be an arbitrary subset of the set of all outcomes, and this allows for interdependencies.

bard's sense is not an empty concept; it means that the right is usually respected.

I conjecture that the rights that were earlier unequivocally characterized as inalienable will never be waived. This is a denial of unrestricted domain, but the class of preference profiles excluded is a narrow one, and there are, I argue, good reasons for believing that people's preferences do not belong to this class. Consider the example about freedom of speech. It is logically possible that everybody prefers that i be deprived of the freedom of speech, but it seems unlikely. This is the type of preference profile I suggest be eliminated from consideration.[31] If I conclude that the rights in the life-and-death example are inalienable (and hence unwaivable), this approach will not work. In the preference profile of the example, which seems reasonable and should not be excluded from the domain, i's right over O and I will be waived according to Gibbard's definition.

The process described in Section II has been brought to an end and a consistent set of conditions has been obtained. (At least, the end has been reached so far as the liberal paradox is concerned. Perhaps I want to impose additional conditions. Then a new round of consistency checks and reconciliation is needed, but I shall ignore this possibility here.) My final, consistent set of conditions is essentially that of Gibbard, but with some restrictions put on the domain of the social decision function, guaranteeing that certain inalienable rights never be waived.[32] In order to use these conditions, I must identify the issues and specify the initial system of rights. Doing so requires information about the particular society under consideration. When all this is done, I am ready to evaluate the social decision function. Again, I must emphasize that the solution is not appropriate if rights are considered inalienable in cases like the life-and-death example; then Pareto optimality has to be weakened.

Has social choice theory really contributed anything to the discussion above, or could I have reached essentially the same conclusion

[31] I should have liked to give a precise statement of the domain restrictions necessary to guarantee that a right never be waived. Since I am not able to do so, the whole discussion must stand as a conjecture.

[32] This should not be construed as an endorsement of every detail in Gibbard's solution. The general idea that a right should be waived when preferences are such that it conflicts with Pareto optimality and other people's rights, can be made precise in several ways. It is not clear that Gibbard's definitions are the most appropriate in every respect.

through purely informal reasoning? Perhaps I could, but it seems unlikely. The contradiction between Pareto optimality and liberalism can obviously be perceived without having been formulated and proved as a theorem. But it is not so easy to see how much the conditions must be weakened. Gibbard's idea of occasionally waiving rights seems natural, or even obvious. It is not equally clear how this idea can be made precise in such a way that all contradictions disappear. In the absence of the formal apparatus of the theory, I believe we would have difficulties in determining whether the waiving of rights according to some specific rule is sufficient to achieve consistency. We might end up having doubts about a system that actually is consistent, or, more likely and more dangerously, we might be able to convince ourselves that an inconsistent system is free of contradictions. The formal theory and its theorems assist us in eliminating these problems.

In order to illustrate the differences between various interpretations of social choice theory, I shall wind up by looking at the liberal paradox in the light of an interpretation different from the one used so far.

Often, the theory is thought of as a tool for constructing decision procedures. Unrestricted domain and collective rationality cannot then be considered consequences of the definitions, as was the case above. If these conditions are to hold, we must make them hold. That is, we must see to it that the decision procedure that is being constructed works for all preference profiles and always produces a non-empty choice set.[33] In the previous discussion, we identified one acceptable reason for violating unrestricted domain; it is unnecessary to insist on this condition if we are able to conclude, on theoretical or empirical grounds, that some preference profiles cannot occur. This reason still applies. In addition, we can decide that certain profiles, though possible, shall not be accepted. In that case, people are not guaranteed that their true preferences can be transmitted to the procedure and taken into account; a person shall have to choose from a more restricted set of preference orderings. This is clearly not a desirable situation, but it may be a lesser evil than violating any other condition.

If we decide that the two conditions mentioned so far have to hold,

[33] Earlier, a multi-member choice set caused no problems. Now we have to end up with one, final decision. That is, we must construct a method for choosing one outcome from the choice set if it has more than one element. Drawing lots is a possibility.

which seems most natural, we are left with a conflict between Pareto optimality and liberalism. In the present interpretation, liberalism means that decision-making is decentralized. Hence the content of the theorem is that decentralized decision-making can produce non-optimal outcomes. This is nothing new; in connection with decisions about the provision of public goods, it has been known for a long time and is referred to as the free rider problem.[34] But the theorem goes further. It tells us that even the tiniest amount of decentralization, namely giving each of two persons control over the choice from one pair of possible decisions, makes it impossible to guarantee Pareto optimality. Those who construct decision procedures have to choose between decentralization and optimality. This is the significance of the liberal paradox in this interpretation.

[34] A well-known presentation of the problem is that given by Samuelson (1954). According to Bohm (1979), a description was given by Knut Wicksell as early as 1896.

REFERENCES

Arrow, K. J. (1963) *Social Choice and Individual Values*, 2nd edn, New York: John Wiley and Sons (1st edn published 1951).

Blair, D. H., Bordes, G., Kelly, J. S. and Suzumura, K. (1976) 'Impossibility theorems without collective rationality', *Journal of Economic Theory* 13, 361–79.

Bohm, P. (1979) 'Estimating willingness to pay: why and how?' *Scandinavian Journal of Economics* 81, 142–53.

Gibbard, A. (1974) 'A Pareto-consistent libertarian claim', *Journal of Economic Theory* 7, 388–410.

Hylland, Aa. (forthcoming) *The Limits of Social Choice Theory*. To be published by Cambridge University Press and Universitetsforlaget in the present series: 'Studies in Rationality and Social Change.'

Kelly, J. S. (1978) *Arrow Impossibility Theorems*, New York: Academic Press.

Kemp, M. C. and Ng Y.-K. (1976) 'On the existence of social welfare functions, social orderings and social decision functions', *Economica* 43, 59–66.

Parks, R. P. (1976) 'An impossibility theorem for fixed preferences: a dictatorial Bergson–Samuelson welfare function', *Review of Economic Studies* 43, 447–50.

Rawls, J. (1971) *A Theory of Justice*, Cambridge, Massachusetts: Harvard University Press.

Roberts, K. W. S. (1980) 'Social choice theory: the single-profile and multi-profile approaches', *Review of Economic Studies* 47, 441–50.

Samuelson, P. A. (1954) 'The pure theory of public expenditure', *Review of Economics and Statistics* 36, 387–9.

Samuelson, P. A. (1967) 'Arrow's mathematical politics', in S. Hook (ed.), *Human Values and Economic Policy*, New York: New York University Press, pp. 41–51.

Sen, A. K. (1970) 'The impossibility of a Paretian liberal', *Journal of Political Economy* 78, 152–7.

Sen, A. K. (1976) 'Liberty, unanimity and rights', *Economica* 43, 217–45.

Sen, A. K. (1977) 'Social choice theory: a re-examination', *Econometrica* 45, 53–89.

3. Laundering preferences*

ROBERT E. GOODIN

Want-regarding moralities are continually embarrassed by the fact that some preferences are so awfully perverse as to forfeit any right to our respect. The latest horror story comes from Sen (1979a, pp. 547–8):

> Consider a set of three social states, x, y and z, with the following utility numbers for persons 1 and 2 (there are no others).
>
> | | x | y | z |
> |----------|-----|-----|-----|
> | person 1 | 4 | 7 | 7 |
> | person 2 | 10 | 8 | 8 |
>
> In x person 1 is hungry while 2 is eating a great deal. In y person 2 has been made to surrender a part of his food supply to 1. While 2 is made worse off, 1 gets more utility, and the sum total of utility happens to be larger (with diminishing marginal utility). . . . Consider now z. Here person 1 is still just as hungry as in x, and person 2 is also eating just as much. However, person 1, who is a sadist, is now permitted to torture 2, who—alas—is not a masochist. So 2 does suffer, but resilient as he is, his suffering is less than the utility gain of the wild-eyed 1. The utility numbers in z being exactly the same as in y, welfarism requires that if y is preferred to x, then so must z be.

That, Sen supposes, is 'what's wrong with welfare economics'. Its 'welfarism' insists that we rank social states on the basis of 'utility

* Earlier drafts benefited from discussion in the ECPR Workshop in Brussels and the Working Group on Rationality at Ustaoset, and especially from the comments of Jon Elster, Dagfinn Føllesdal, Aanund Hylland, Herman van Gunsteren, Francis Sejersted and Mike Taylor.

information' alone. Where such perverse preferences are involved, we are otherwise inclined. We want to bring 'non-utility information' to bear on the social choice, most especially in the form of vested rights guarantees protecting people from the meddlesome (or, indeed, sadistic) preferences of others (Sen 1970, Ch. 6 and 6*; Sen 1976).

In so arguing, Sen—like many others—beats too hasty a retreat. The answer he seeks may yet be found 'in the shadow of utilitarianism' (Hart 1979, p. 98) if only we look hard enough.[1] The problem, basically, is that preferences sometimes seem 'dirty'. Surely it makes sense to see whether they cannot somehow be 'laundered' before we discard them altogether. The argument of this essay is that we hesitate to launder preferences only because we are unsure of their fabric.

Recourse to 'non-utility information' seems necessary merely because we work with such an impoverished conception of individual preferences in the first place. For the most part, they are just taken to be an individual's ranking of various social states. Whatever underlies this ordering ordinarily goes undiscussed (Lancaster 1967). But, in truth, there is much more to individual utilities than is captured by simple numbers and rank-orderings.[2] 'Utility information' can and should be seen to include information about *why* individuals want what they want, about the *other things* they also want, about the *interconnections between* and *implications of* their various desires, etc. Obviously, this goes well beyond the sort of information social choice

---

[1] In another paper, Sen (1980–1, pp. 193–4, 207, 210) acknowledges that a wider range of '"utility-supported" moralities permits evaluation of states of affairs distinguishing between different components of each person's utility, and possibly weighting them differently'. These moralities are 'utility-supported' in so far as they 'justify all moral values with reference to some aspect of utility'. But in them 'the question of weighting of different types of pleasures is left open as an additional moral exercise'—drawing (presumably) upon non-utility considerations. The present discussion, in contrast, looks for internal (i.e. preference-based) justifications for laundering preferences.

[2] Sen himself, in an earlier essay (1977, pp. 335–6), had complained similarly that 'traditional [utility] theory has *too little* structure. A person is given *one* preference ordering, and as and when the need arises this is supposed to reflect his interests, represent his welfare, summarize his idea of what should be done, and describe his actual choices and behaviour. Can one preference ordering do all these things? A person thus described may be "rational" in the limited sense of revealing no inconsistencies in his choice behaviour but if he has no use for these distinctions between quite different concepts, he must be a bit of a fool. . . . Economic theory has been much preoccupied with this rational fool decked in the glory of his *one* all-purpose preference ordering. To make room for the different concepts related to his behaviour we need a more elaborate structure.'

theorists ordinarily ask us to collect—or their models are capable of processing.[3] But that does not make it non-utility information.[4] The information in question is still very much information about individual utilities.

The ultimate goal of enriching our utility information in this way is to use it to launder people's preferences. The thin theory of individual preferences leaves opponents of want-regarding moralities too easy a task: they need only nod in the direction of some incredibly nasty preferences and say, 'There must be something terribly wrong with any principle that requires us to respect *those* preferences.' Want-regarding moralities can be spared that sort of pre-emptory dismissal by showing that there are reasons, internal to preferences themselves, for disregarding some sorts of preferences.

The plan of attack is as follows. Section I argues that 'censoring' utility functions is a more adequate response to the concerns that drive us to such non-utility recourses as the ascription of rights. Section II reveals various ways in which laundering preferences could, in principle, be perfectly consistent with respecting preferences construed more broadly. Section III goes on to argue that, in the context of collective decision-making, people are forced to undertake a limited laundering of their own preferences. And Section IV discusses practical devices for such further laundering of people's preferences as Section II might warrant.

I. Input versus output filters

Allowing social decisions to turn strictly on individual preferences might, in communities of sufficiently bloody-minded individuals, produce some pretty onerous outcomes. Surely it is inadequate to fall back (*pace* Smart 1973, pp. 67–73) on the purely contingent proposition that vicious preferences are uncommon. By now, 'it would be common ground to nearly all supporters of democracy that there are

---

[3] 'If social choice were to depend not merely on individual preferences but also on other things, e.g. the causation of these preferences, then the concept of a collective choice rule . . . is itself in doubt' (Sen 1970, p. 85).

[4] Cf. Sen (1979b, pp. 482–3): 'Non-utility information relating to *how* "personal" choices are, what *motivation* the persons have behind their utility rankings, whether the interdependence arises from liking or disliking the others' physical *acts* . . . or from the *joys and sufferings* of others, etc., may well be found to be relevant in deciding which way to resolve the conflict. If so, then the adequacy of utility information is denied.'

certain laws or regulations that ought not be passed even if the greater part or indeed the whole of the people favour them' (Wollheim 1958). In this section, I shall compare alternative methods of imposing and justifying such restrictions on collective choice.

The fundamental contrast is between strategies which filter *outputs* of a social decision function and those which filter *inputs* into it. Whereas output filters work by removing certain options from social consideration, whatever their utility, input filters work by refusing to count certain classes of desires and preferences when aggregating individual utilities. Output filters can be conceptualized as barriers erected at the back end of the social decision machinery, preventing policies predicated on perverse preferences from ever emerging as settled social choices. Input filters might be regarded as barriers erected at the front end of the social decision machinery, preventing perverse preferences from ever entering into consideration.[5] Or in more formal terms: output filters act as 'stopping modals' telling us 'you can't do that' (Anscombe 1978); input filters act to provide 'exclusionary reasons' banning consideration of certain sorts of reason for action altogether (Raz 1975, Ch. 1.2).

The threat of meddlesome preferences ordinarily drives us to strongly anti-utilitarian recourses such as vested rights. That classically liberal democratic response is, with the shift 'between utility and rights' (Hart 1979), once again in philosophical favour. The hope is to avoid heinous outcomes by ascribing to each individual a set of rights, thereby circumscribing the application of utility reckoning. Rights function as a ' "No Trespassing" sign, . . . a fence erected around an area from which the majority would be excluded by constitutional law' (Mayo 1960, p. 188). They create a 'protected sphere' (Hayek 1960) and guarantee individuals 'protected choices' (Hart 1955).

All those phrases seem to suggest that vested rights characteristically act to filter outputs rather than inputs. This emerges especially

---

[5] Input filters generally tend to come into play earlier (and output filters later) in the social decision process. But there is more to the distinction than that. An output filter taking effect at the earliest possible moment, excluding certain outcomes from the feasible set (and hence from further consideration) right from the outset would not be equivalent to an input filter: that would still allow people's perverse preferences to shape their responses to the remaining options, in a way that input filters would preclude; and, besides, excluding all options that could conceivably evoke people's perverse preferences prevents them from bringing non-perverse preferences to bear on those options, in a way that input filters would not.

clearly in the 'general theory of rights' offered by Dworkin, who similarly sees the problem with want-regarding moralities as residing in the meddlesomeness of 'external preferences':

> The concept of an individual political right . . . allows us to enjoy the institutions of political democracy, which enforce overall or unrefined utilitarianism, and yet protect the fundamental right of citizens to equal concern and respect by prohibiting decisions that seem, antecedently, likely to have been reached by virtue of the external components of the preference democracy reveals (1977, p. 277; cf. 1978, pp. 134–5).

On Dworkin's account, rights prevent certain kinds of decisions from emerging *out* of the social calculus, rather than preventing offensive 'external preferences' from entering *into* it. And this is, in practice, just how most (if not all) rights work. For the most part, rights restrict results or, at most, procedures. Only very occasionally do they restrict inputs *per se*.[6] Hence, the standard response to the problem of perverse preferences is essentially a variant of the output-filtering strategy.

The first problem with the rights strategy is how to circumscribe and justify the creation of this private preserve. There is, of course, the familiar proposition that the state has no business interfering with 'purely private-regarding' actions. But if that means actions affecting no one else, Stephen (1874) was right to object that every action has an impact upon someone else besides the actor—there are simply no private-regarding actions in that sense.[7] Nor did Mill (1859) claim there were. Rather, he defined private-regarding acts as those not impinging upon anyone else's interests, 'or rather certain interests, which, either by express legal provision or tacit understanding, ought

---

[6] One example might be that American courts suppress evidence obtained in violation of defendants' constitutional rights from their trials. But even this is the consequence not of rights *per se* but rather of the particular strategy (the 'exclusionary rule') chosen to enforce those constitutional constraints. There is no logical reason why your right against random searches need strictly entail a right to have illegally-seized (but undeniably true) evidence excluded at your trial.

[7] We could, of course, give a narrow meaning to the notion of 'being affected' as e.g., 'having one's life materially impinged upon' (Barry 1965, p. 63). But there is no reason to protect people's material interests to the exclusion of their equally or more important symbolic ones in protecting their self-respect, for example (Goodin 1980, Ch. 5; 1982, Ch. 5).

to be considered as *rights*'. He goes on to say that 'the acts of an individual may be hurtful to others, or wanting in due consideration for their welfare, without going to the length of violating any of their constituted rights', in which case the offender may not legitimately be punished by law (Mill 1859, Ch. 4; Gray 1981, pp. 98–101). Thus, the private-regarding move will not suffice to justify the ascription of rights: either we take 'private-regarding' literally, in which case the private sphere is the empty set; or else we define the private-regarding sphere in terms of rights, which makes it circular for us then to justify rights in terms of the private-regardingness of actions within that sphere.

A more plausible argument for carving out a private sphere builds upon the notion of 'respect for persons' (Fried 1978, pp. 28–9). If we follow Benn (1971, p. 8) in conceiving of a person as being essentially 'a subject with consciousness of himself as agent, one who is capable of having projects, and assessing his achievements in relation to them', then to respect someone as a person we must allow him free rein (within limits, perhaps) to frame and pursue projects as he will. Rights are ascribed to individuals on this reading of the situation, in order to create and protect this private sphere within which people will be free to engage in that distinctively human activity.

This however, raises a second problem for the rights strategy. If we are truly concerned to show people respect, we must not confine ourselves (as the rights strategy does) to prohibiting degrading policy outcomes. We also show people respect or disrespect through our attitudes and motives, even if they do not culminate in actions (Goodin 1982, Ch. 5). Liberals count all preferences without prejudice —humiliating and degrading ones included—in the social decision calculus. Vested rights guarantees, by filtering some nasty options out of the feasible set, might save the state from *doing* anything that shows citizens disrespect. It has, however, already shown them disrespect by *counting* degrading preferences in the first place. Input filters will be required if we are to prevent the sort of humiliation that comes from the social sanctioning of mean motives.[8]

---

[8] Output filters might usefully *supplement* input filters, serving as a second line of defence in case anything slips through the first. For purposes of protecting self-respect, however, they can never *substitute* fully for them (*pace* Dworkin 1977, Ch. 12; 1978, pp. 134–5).

Vested rights have been used as a practical illustration of the output-filtering strategy at work. There are few (if any) mechanisms already in operation approximating the input-filtering ideal. But as a theoretical option this strategy is familiar enough. There have been regular suggestions that we 'censor utility functions' (Harsanyi 1977, p. 62) in one way or another: according to Rousseau's (1762) dictum, people's 'particular wills' should be excluded in reckoning the 'general will'; according to some economists, we should weight 'merit goods' more heavily than preferences warrant (Musgrave 1968; Head 1974, Ch. 10 and 11); and according to some political theorists, we should discount preferences which are 'external' in nature (Barry 1965, pp. 62–6; Dworkin 1977, Ch. 9 and 12; 1978, pp. 134–5; 1981, pp. 196–204). Thus, filtering inputs is a live, albeit neglected, theoretical option. And it certainly is one that ought to be pursued since it is a better response to the sorts of concerns that drive us from preference-based moralities to rights-based ones.

II. Grounds for laundering preferences

The great challenge lies, of course, in finding acceptable grounds to justify the laundering of preferences.[9] Here I shall concentrate on the most unobjectionable form, viz., grounds internal to the preferences themselves. Provided we are prepared to consider richer utility information, much laundering can be justified in strictly want-regarding terms. Taking into account why individuals want what they do, what else they want, and the relationships between and implications of their various desires, five especially interesting justifications for censoring utility functions emerge:

1. *Protecting preferences from choices*. A person's choices do not always perfectly reflect his preferences, as Sen (1973; 1977, pp. 327–9)

---

[9] 'Intuitively, of course, we feel that not all the possible preferences which an individual might have ought to count; his preference for matters which are "none of his business" should be irrelevant. Without challenging this view, I should like to emphasize that the decision as to which preferences are relevant and which are not is itself a value judgment and cannot be settled on an *a priori* basis' (Arrow 1963, p. 18). Notice that the artificial constraints on the feasible set imposed by vested rights guarantees must similarly emerge from a prior collective choice which, *ex hypothesi*, is not constrained in these ways.

has long said and several contributors to this volume (especially Gibbard) further emphasize. Sometimes people make their choices on the basis of incomplete information, in ignorance of their own future desires (Goodin 1982, Ch. 3), or in the absence of a 'full and vivid awareness' of all the alternatives (Gibbard infra). At other times, people's choices merely reflect the framework within which they are set, or their desire to avoid responsibility for risks, rather than their preferences among the alternative outcomes themselves (Tversky and Kahneman 1981). In such cases, we can serve a person's 'real' preferences only by censoring the misleading statement of his preferences revealed in his choices.[10]

2. *Reciprocal forbearances.* Much of social morality has now been explained in terms of reciprocal forbearances. Many of the things that would benefit one person would harm another person even more; so each agrees to forgo his opportunities for gain at the other's expense, on condition that the other likewise forgo opportunities for gaining at his expense. The much-discussed strategic problems associated with concluding and enforcing such agreements need not detain us here. We need only note that reciprocal forbearances might apply to preferences as well as to actions. People may reciprocally agree to a rule of mutual tolerance, each forgoing his meddlesome preferences on condition that all others do likewise. Wherever people agree (explicitly or implicitly, actually or hypothetically) to such mutual forbearances, laundering their preferences accordingly is fully justified in terms of their own larger preference ordering.

3. *Explicit preferences for preferences.* On some accounts, people are

---

[10] Surely we find something fishy in the idea of *forcing* people to consummate Pareto-optimal deals (Broome 1978, p. 316; Barry infra). But that is merely because 'deals' conventionally refers to a class of market choices we feel should be left to individuals themselves to botch as they will. (Why we have these feelings is an interesting question, but irrelevant for present purposes.) Where we really are trying to make social choices by aggregating people's preferences, the situation is very different indeed. There would be something even *more* fishy about feeding into those collective deliberations the misleading preferences revealed by people's suboptimal choices, assuming some other truer readings of their preferences can be obtained. This might be done either directly (by asking them) or indirectly (inferring preferences from other choices, as Tversky and Kahneman's [1981] experiments do).

distinguished from lower forms of life precisely by the fact that they have not only preferences but also preferences for preferences (Frankfurt 1971). These may be moral ideals (Sen 1974), personal ideals, social ideals or role preferences (Goodin 1975; Benn 1976, 1979; Hollis 1977). Whatever the details, such preferences can justify censoring utility functions of those possessing them whenever, through 'weakness of will', people fail to live up to their second-order preferences. Frequently, such individuals find that one set of preferences actually guides their behaviour while they dearly wish another would instead. Laundering their preferences then simply amounts to respecting their own preferences for preferences.[11] In aggregating preferences, we count only those the individual wishes he had; and we ignore all those he wishes he did not actually experience. No violation of want-regarding principles is entailed, since we choose which preferences to count and which to ignore strictly on the basis of the individual's own (higher order) preferences.

4. *Implicit preference for preferences.* People only occasionally acknowledge explicitly their preferences for preferences. We can, however, often find logical relations among those preferences that people *do* acknowledge which imply certain other preferences that they may fail to acknowledge. Admit it or not, these are still *their* preferences in some important sense. Just as explicit preferences for preferences can justify ignoring certain first-order preferences in conflict with them, so too might certain logically central implicit preferences override (and justify decision-makers in ignoring) certain explicit but logically more tangential preferences. The social decision-maker would be ignoring someone's first-order preference so as better to serve his own higher-order (albeit implicit) preference. Preferences

[11] Rothenberg (1968, p. 330) similarly argues that the preferences of consumers can be 'corrected' without violating their sovereignty because 'consumers have a hierarchy of wants'. On a crude Humean philosophy of the mind, where desires are distinguished only according to their intensity, we may well wonder why a weak second-order preference should override a stronger first-order one (Hollis 1981). But cashing everything out in terms of undifferentiated desires is just the philosophical precursor to the fallacy that is here criticized, viz., that everything can be fitted into a single, undifferentiated preference ordering. Once we realize that people are self-conscious agents, and that some preferences crucially confirm their self-conceptions while others tend to compromise them, we find grounds for respecting preferences for preferences (which reflect those self-images) in lieu of choices *per se* (Goodin 1982, Ch. 5).

still dictate decisions—no extraneous considerations are being introduced—so welfarism and want-regarding principles are not violated.[12]

Many weakly paternalistic arguments appeal in precisely this way to implicit preferences for preferences to justify revising people's statements of their preferences. Gerald Dworkin (1971, pp. 120–1) asks us to:

> suppose that there are 'goods' such as health which any person would want to have in order to pursue his own good—no matter how that good is conceived. . . . Then one could agree that the attainment of such goods should be prompted even when not recognized to be such, at the moment, by the individuals concerned.

The example he offers concerns an automobile driver who, although fully and vividly aware of the risks he runs on the roads, refuses to fasten his seatbelt. 'Given his life-plans, . . . his interests and commitments already undertaken', Dworkin writes, 'I think it is safe to predict that we can find inconsistencies in his calculations at some point'; and, therefore, we can use his deeper (albeit only implicit) preference to justify disregarding his stated preference for not bothering to belt up. Or, as another example, Rothenberg (1968, p. 330) argues that 'drug addicts are deemed not to be able to act in their own best interests'—and we are justified in ignoring their demand for heroin—because that preference 'is inconsistent with the fundamental pattern of their system of ends' and the deeper preferences implicit within it.

Another example builds on the 'master–slave paradox', posed by Elster (1976, p. 261) as follows:

> The master is caught in a trap, for he can get no real satisfaction from his power over a being that he treats like a thing. The very concept of unilateral recognition is contradictory, as can be seen by thinking through the farcical idea of a nation being diplomatically recognized

---

[12] Dworkin's (1977, Ch. 7) argument for 'taking rights seriously' is similar: given the rights which the Constitution does guarantee explicitly, there are further rights which, although not explicity recognized, are nevertheless logically implicit in the set of rights which *are*; and these implicit rights not only must be considered alongside those which are enumerated but also, being logically more central, must typically *override* them in instances of conflict.

by one of its own colonies. To the extent that the master treats the slave on a par with cattle, he gets no non-economic satisfaction from his power; to the extent that he treats the slave like a human being, he has no power over him.

From this, we can generalize important conclusions about the whole class of 'second-party preferences', defined broadly as preferences about other people's satisfactions whether of a benevolent or malevolent bent. One can find it satisfying to have and act upon such preferences about other people only if they are acknowledged to *be* people, either by one's own behaviour or by some larger system of social values. For the altruist this creates no problems. But for the sadist it does: both his personal dealings and his first-order preferences tend to degrade and humiliate people, to reduce men to the level of beasts or worse; if he gets his way on policy, however, he will find it no more satisfying to kick a man than a dog, since the social recognition of a man's greater dignity was all that made kicking him more satisfying in the first place. The sadist's preferences for humiliating men in everyday life, then, imply a preference that others' dignity be acknowledged and respected in public policy. Even though he insistently demands the contrary, the social decision-maker is obliged to respect this logically central implication of the sadist's first-order preferences. His utilities are still controlling, however. No violation of want-regarding principles has occurred.

5. *Internal logic of preference aggregation.* All the previous arguments justify ignoring some of a person's preferences in terms that are somehow internal to his own preference function. A final class of justification points instead to reasons which are located in the logic of the social decision process. Elsewhere I have shown how one such implicit commitment constrains social decisions: respecting people's choices implicitly commits us to respecting their dignity, and this implies certain clear limitations on the choices we may be obliged to respect (Goodin 1982, Ch. 5). Similarly, our very choice of aggregating preferences as a way of making social decisions carries consequences for the kind of preferences that we can count. 'After all', Harsanyi (1977, p. 62) reflects:

the entire basis for [the social decision-maker] i's interest in satisfying j's preference is human sympathy. But human sympathy can hardly impose on i the obligation to respect j's preferences in cases where the latter are in clear conflict with human sympathy. For example, human sympathy can hardly require that i should help sadists to cause unnecessary human suffering—even if a very large number of sadists could obtain a very high utility from this activity.

Just as some aspects of our friends' characters are better ignored, so too are some arguments in our fellow citizens' preference functions better ignored.[13]

Taken together, these five arguments suggest that want-regarding principles, suitably expanded, can afford substantial protection against perverse, meddlesome preferences and the heinous outcomes they might produce. Ultimately we may want to launder preferences more thoroughly than we can find internal justifications (like 1–5 above) for doing, and we will be forced to fall back upon ideal-regarding principles to do so (Barry 1965, p. 40). But at least I hope to have shown that the demand for decisions based on 'non-utility information'—and on vested rights in particular—is premature. There is still much unexploited room for manoeuvre within the want-regarding framework, provided we take a broad view of what 'utility information' can include.

III. The self-laundering of preferences in collective choice

Formal models of collective choice tend to represent it as some mechanical process of aggregating individual preferences. This badly understates the true complexity of the process. Whereas these models usually take preferences as given, for example, classical theories of democracy have always acknowledged that people can and should reformulate their preferences in response to rational discussions in the

---

[13] J. S. Mill (1859, Ch. 5), who is otherwise prepared to respect any self-regarding choice a person might make, offers a parallel argument against allowing a person to sell himself into slavery: 'The principle of freedom cannot require that he should be free not to be free. It is not freedom to be allowed to alienate his freedom.' Notice that such arguments do not exclude anyone from the decision arena. Everyone's preferences count, but not *all* of each person's preferences necessarily count (Harsanyi 1977, p. 62).

course of collective deliberations: instead of working on some *fixed* set of preferences, the social decision machinery changes them in the process of aggregating them.[14] Furthermore, and more to my present point, the social decision machine does not necessarily work with each individual's *full* set of preferences. This is partly because people find strategic advantage in suppressing some of their preferences. But those distortions are ephemeral, shifting with strategic circumstances. There is a deeper dynamic, inherent in the very nature of the collective decision process, which induces people to launder systematically their own preferences, and to express only a small subset of their preferences in the form of political demands. This forms the basis for my discussion here.

Basically, collective decision-making evokes different sorts of preferences, because 'an individual's response depends on the institutional environment in which the question is asked'. Certain kinds of argument, powerful though they may be in private deliberations, simply cannot be put in a public forum. In countries like the United States,

the market is an institution designed to elicit privately-oriented responses from individuals and to relate those responses to each other. . . . The electoral, legislative and administrative processes together constitute the institution designed to elicit community-oriented responses (Maass 1966, pp. 216–17; cf. Musgrave 1968, p. 160).

This latter function is served even more dramatically, in both theory and in practice, by Thai village meetings. Anthropological observers report substantial discrepancies between the way in which villagers say they intend to vote on public works projects ahead of time, and the way they actually do vote in the end. Closer examination of what happens during collective deliberations leading up to these votes reveals that

---

[14] Little (1952, p. 428) says that we can 'sidestep' this problem, 'formally at least', by allowing 'as much revoting, or rearrangement of individual [preference] orders, as the individuals want when they learn the opinions and verdicts of each other. Accordingly, each order is, so to speak, a deadlock order. It is the individual order resulting when all persuasion has been used, and after an indefinite number of straw votes.' Focusing exclusively upon people's 'deadlock orderings', however, loses track of the *process* by which the mutual influence was accomplished, and this is what is ethically crucial. 'Persuasion', where everything is above board, is perfectly proper; devious 'manipulation' or 'demand-management' is not (Goodin 1980; Dunleavy and Ward 1981).

'during the course of the decision period the community-oriented values become increasingly important' (Bilmes 1979, p. 174).

The distinction I am drawing is not between a person's 'perceived' interests and some other 'truer' interests which he discovers in the course of collective decision-making.[15] Instead, it is between multiple preference orderings actually operative within the individual, and which he applies differently according to the context. As Brandt (1967, p. 27) emphasizes in his reply to Arrow, 'some choices are motivated by the prospect of enhanced *personal welfare* . . . whereas others'— paying taxes is his example—'are motivated by considerations of *moral principle*'.[16] People can be moved to action by either self-interest or benevolence and 'moral sentiments' (Hume 1739, 1777; Smith 1790), by 'ethical' as well as egoistic preferences (Harsanyi 1955; Goodin and Roberts 1975). Partially overlapping this category of 'ethical preferences' is that of 'social' as distinct from 'private' preferences (Plamenatz 1973, pp. 155–68), or what Bator (1957) calls people's 'welfare function' as distinct from their private 'utility functions'.[17]

This multiplicity of preference orderings matters because, in the context of collective decision-making, people will launder their own preferences. They will express only their public-oriented, ethical preferences, while suppressing their private-oriented, egoistic ones. There are various reasons for this. One explanation turns on the logic

[15] Thus, my distinction has nothing to do with what Arrow (1963, pp. 82–3) calls Kant's 'idealist' doctrine that 'each individual has two [preference] orderings, one which governs him in his everyday actions and one which would be relevant under some ideal conditions and which is in some sense truer than the first ordering'.

[16] Of course, moral principles themselves can be represented simply as constrained egoism. Runciman and Sen (1965, p. 557; cf. Arrow 1967) write that 'our interpretation . . . does not require us to impute to each person more than a single set of orderings. . . . The conflict between the [egoistic] will of all and the [altruistic] general will arises not because the individual must be required to change his preference orderings, but because of the difference between the outcome of individual strategy and of enforced collusion which arises under the conditions of the non-cooperative, non-zero-sum game.' Thus, players 'cooperate' in Prisoner's Dilemma games not out of moral principles but only out of enlightened self-interest. No doubt this is part of the story, but only part of it (Goodin 1982, Ch. 6).

[17] See, similarly, MacRae 1976, pp. 138–45; Self 1975, pp. 134–5; Benn 1979; Margolis 1981. All these different sorts of preferences can, of course, be represented in a single meta-ordering, provided the person invokes each set in some systematic and consistent fashion (Benn 1979, p. 302; Brandt 1967; Sen 1977). My argument is not that it is impossible to do so, but merely that it is inadvisable—too much useful information gets lost through such representations.

of 'role rationality' (Benn 1979; Hollis 1977; Goodin 1975). Publicly-oriented preferences are somehow uniquely appropriate to the role of the individual *qua* citizen. An individual can hardly conceive of himself as a person who occupies a certain kind of role, and who has the sort of preferences that go with it, without also being disposed to act upon them when required by the role to do so (Benn 1979).

Paralleling this argument for moral self-expression couched in terms of roles and self-images is another even stronger argument growing out of the so-called 'paradox of voting'. The 'paradox' is that anyone bothers to vote at all, since the chance of his casting the decisive vote 'is less than the chance he will be killed on the way to the polls' (Skinner 1948, p. 265; Downs 1957, Ch. 14). That certainly makes it irrational to vote, in so far as your payoffs depend upon the outcome of the election. Privately-oriented egoistic preferences are of that sort and, therefore, cannot motivate your vote. Ethical preferences, however, are not necessarily of that sort. 'Doing the right thing' may bring satisfaction in itself, even if the ultimate outcome is unchanged. Repudiating some nasty business can enhance your self-respect and sense of integrity, even if it goes ahead without your endorsement.[18] Why people bother going to the polls at all, and how they vote once they get there, can both be explained in terms of taking a moral stand.[19]

[18] See, e.g., Williams 1973, pp. 108–18; Hill 1979; Goodin 1982, Ch. 5. Since one vote, taken by itself, has virtually no consequences, electoral or otherwise, voters are precluded from acting on consequentialistic principles for consequentialistic reasons. This may lead people to behave more fanatically in electoral contexts than in personal dealings, where due regard for the consequences may make them back off their more brutal principles—thus Hollis (1981) is led to dub the man of principle 'Nasty'. But it is also possible for people to embrace consequentialistic ethics as a matter of *principle*. A World War II pilot, for example, might have refused to participate in the firebombing of Dresden because he regarded it as immoral on strictly consequentialistic grounds (100,000 would die without advancing Allied military objectives one bit), even though he recognized that his refusal to participate itself carried no consequences (there were plenty of other pilots willing to take his place).

[19] Downs (1957, Ch. 14), seeing the paradox, suggested that people go to vote out of a feeling of 'civic duty' but that, once they get to the polling station, self-interest once again seizes them. Downs was roundly criticized for introducing this *deus ex machina* to save his theory. Here I hope to have removed some of the sense that this is an utterly ad hoc improvisation by deriving it from some other plausible assumptions about the nature of preference functions, and by putting the proposition to wider use than does Downs himself. (For further analysis, see Goodin and Roberts 1975). This argument might also explain the tendency towards convergence Downs observes in political competition. The ethical tastes that collective decision-making evokes are likely to be—and, on some analyses, must necessarily be—less diverse than the egoistic desires that drive individuals' more ordinary market decisions.

The power of a moral crusade to draw votes is well known to politicians themselves. Academic commentators have been slow to pursue the point. It has long been thought a topic better left to crowd psychologists. Among political scientists, the possibility of a fundamentally ethical 'public-regarding' ethos underlying voting behaviour attracted little attention after Banfield and Wilson's (1964) seminal study.[20] Recent work, however, provides at least indirectly support for this proposition. Sears et al. (1980), studying policy positions and presidential voting among the American electorate, find that 'symbolic values' are far more powerful predictors than 'self-interest'. And Kinder and Kiewiet (1981) show that, in so far as economics dictates votes at all, people respond more strongly to changes in general economic conditions nationwide than to changes in their own personal finances. The explanation I would offer for both sets of findings (although not that offered by the authors themselves) is that the 'paradox of voting' has forced voters to take a morally principled rather than a narrowly self-interested stand.

For a practical illustration of citizens using different sorts of preferences in their private and public capacities, consider the problem of time-discounting. One of the most compelling reasons for an individual to discount future costs and benefits relative to present ones is that, come the time, he may not be alive to experience them. How heavily he should discount payoffs depends upon how far they are in the future and, of course, upon his own life expectancy. The risk of death would, according to mid-century actuarial statistics, imply a discount rate of around 0.4 per cent for a 40-year-old American; for an Indian of the same age, it would be 2.15 per cent (Eckstein 1961, p. 457). From a purely individual point of view, it would be foolhardy *not* to discount future payoffs by at least this much.[21] But risk-of-death discount rates have no place in social decision-making. 'Mortality probabilities of specific individuals become irrelevant' there, because 'the society goes on forever' (Eckstein 1961, p. 459).

[20] Flawed as their operationalization and evidence may be, Banfield and Wilson's (1964) conclusion that the rich are more public-regarding should come as no surprise. Assuming everyone has exactly the same ethical and egoistic preferences, and assuming that both display diminishing marginal utility, those who are materially better off will naturally attach relatively more importance to their ethical goals (Margolis 1981).

[21] Similarly, your values and preferences can change so much over time that your former self has, in effect, expired, which would increase the risk-of-death discount rate even further (Parfit 1976, pp. 98–9).

Eckstein, like Pigou (1932, Pt. I, Ch. 2) before him, thinks this provides the basis for 'dictatorial' action on the part of social planners in overriding citizens who try to impose their inappropriately personal discount rates on social undertakings. In practice, however, this seldom seems necessary. What Pigou and Eckstein overlook is the possibility, broached by Marglin (1963, p. 98), that 'an individual's time-preference map may be strongly influenced by his expected life span in his capacity as an economic man, although in his role as a citizen his time-preference map may reflect great concern for adequate protection for posterity'. When considering social projects, people do seem to take a far longer view than would make sense in their private lives, voting for projects that would hardly begin to pay off until they have long been in their graves. The reason, I suggest, is that in the context of collective decision-making people operate on the basis of social/ethical preference functions rather than upon their private/egoistic ones.[22]

IV. Beyond self-laundering

The laundering of preferences is easily accomplished—and even more easily justified—when in the context of collective decision-making people do it themselves.[23] But some perverse preferences ('principled malevolence', for example) would still slip through. Such further laundering as is required must be undertaken by policy-makers, acting on behalf of those with whom they interfere. This paternalism is gentled by two special features. First, overriding a person's preferences is (on arguments 1–4, at least) justified in terms of that person's *actual* preferences, rather than in terms of some reading of his 'true

[22] 'People of underdeveloped countries' such as India do, as Eckstein (1961, p. 459) says, 'place a considerable premium upon benefits in the early years of development projects.' But that can be explained, as in footnote 19 above, by saying that very poor people place very little weight on the ethical and publicly-oriented components of their preference functions.

[23] Before blindly endorsing people's choices (here, to launder their own preferences), we must enquire into the conditions giving rise to those choices. In this case, the constraints seem morally appropriate, arising as they do merely from the fact that each person must blend his own will with others' in reaching a collective decision. We would, however, have no reason to respect choices immoral constraints might force upon people. Thus, it is not enough to say that people choose to launder their own preferences—to justify the laundering, we must add that the circumstances compelling them to do so are morally acceptable.

interests' which he does not himself share. And, second, this argument suggests only that we *disregard* certain votes a person might cast or demands he might make, rather than heavy-handedly force him to do or restrain him from doing something contrary to his will.

The practical political problems surrounding any form of paternalism remain. We need institutions with enough slack between citizen demands and social decisions to allow policy-makers to override mistaken preferences when necessary, but not so much as to cut politicians free from the constraints of citizen preferences altogether. These difficulties are lessened where (as in arguments 1–3 in Section II above) people will eventually come to be grateful to a paternalist who has truly rendered them a service. Then democratic accountability— understood as a *post hoc* check—can suffice. This allows public officials to serve the 'public interest', understood as what the public will eventually come to approve, rather than cater to the public's possibly mistaken ideas of what it might like *ex ante*.[24]

Another modest but nonetheless useful method of laundering preferences, which can be used when that one cannot be, is suggested by the analysis in Section I above. There I argued that our paramount goal should be to protect people's self-respect and dignity, and that these are offended by the social sanctioning of mean motives of others that takes place when perverse preferences are allowed to enter the social decision calculus. That is why input filters (such as preference-laundering) rather than mere output filters (such as vested rights) are required. Notice, however that stopping people from having (or even from giving voice to) such preferences is not what is crucial. Rather, it is stopping the social decision-making apparatus from taking official cognizance of them.

The most straightforward way of doing that is through the rules governing debate within the legislature. Already there are certain limitations on what may be said in the chamber. The rules of the British House of Commons, for example, prohibit 'treasonable or seditious language or a disrespectful use of Her Majesty's name' (May 1971, p. 415); and in the American Congress members may not speak 'disparagingly about any state of the Union' or subject the President or other

---

[24] Elaborate institutional reforms along these lines might not be necessary if, as Fiorina (1981) argues, most voting in ordinary elections is 'retrospective' rather than 'prospective' anyway.

members of their own or the other chamber to 'personal abuse, innuendo, or ridicule' (Cannon 1963, p. 157; US Congress 1965, Senate Standing Rule 19.4). Other 'rules of decorum' restrict the form speeches can take. Drawing upon British precedents, Jefferson's *Manual* for the American Congress stipulates, for example, that:

> No person is to use indecent language against the proceedings of the House. . . . No person, in speaking, is to mention a member then present by his name . . . nor to digress from the matter to fall upon the person by speaking, reviling, nipping, or unmannerly words against a particular member. . . . No one is to disturb another in his speech by hissing, coughing, spitting, speaking, or whispering to another, nor stand up to interrupt him; nor to pass between the Speaker and the speaking member, nor to go across the House, or to walk up and down it, or to take books or papers from the table, or to write there (Jefferson 1801, Sec. 17; cf. May 1971, Ch. 19).

At present, these courtesies are extended only to other members.[25] But I can see no reason of principle why they should not be extended to private individuals, groups, corporations, etc.[26]

The effect of extending parliamentary privileges in this way must not be overstated. Banning racist talk from the floor of the chamber will

---

[25] 'It is interesting to note that parliamentary language is not required for personal references to individuals who are not members of Congress. In fact, under the constitutional provision that "for any speech or debate in either House, they shall not be questioned in any other Place" [as under Article 9 of the Bill of Rights for the English Parliament], members are secure against any libel suits. This has led to many dramatic instances of vehement personal attacks upon the integrity and loyalty of various individuals' (Gross 1953, pp. 371–2).

[26] We might be tempted to phrase this as giving citizens a *right* not to be insulted in parliamentary debates. But notice that the parliamentary privileges which are being extended to them do not, strictly speaking, take the form of rights: no member can waive his privilege not to be addressed by name or in insulting language on the floor; nor need a member demand his 'rights' in this regard, the Speaker being charged with the duty of securing order even in the absence of complaints (US Congress 1965, Senate Standing Rule 19.4; Cannon 1963, p. 159; May 1971, pp. 424–33). In the House of Commons, the offending member is required to apologize to the satisfaction of both the House and the offended member; but if the offended party refuses to accept the apology, he is *himself* taken into custody, a practice presumably dating from the days of duels (May 1971, p. 420). Individual rights would ordinarily be expected to display none of these features. Of course, we can always phrase these privileges as 'rights to have certain kinds of inputs restricted'. But they would be very different sorts of rights—operating in different ways and justified on different grounds—than the standard form of rights criticized in Section I.

not prevent measures designed to disadvantage racial minorities from being introduced and passed. At most, my proposal would prevent explicitly racist justifications for them from emerging in the course of parliamentary debates and from entering into the legislative history of the bills eventually enacted. The real motives of their sponsors may be an open secret. Indeed, they may even boast of their racist intentions outside of the chamber itself. The legislature refusing to take official cognizance of perverse (e.g. racist) preferences would, therefore, amount to little more than a polite fiction.[27] That, however, nowise diminishes the value of the gesture in protecting people's self-respect. It is in the nature of dignities and indignities that they are conferred more through symbols and gestures than anything else (Goodin 1980, Ch. 5; Goodin 1982, Ch. 5).

The political realism of this proposal may well be queried. As Dworkin (1978, p. 134) says, it would be futile 'simply to instruct legislators, in some constitutional exhortation, to disregard the external preferences of their constituents. Citizens will vote these preferences in electing their representatives, and a legislator who chooses to ignore them will not survive.' But my proposal does not require politicians to ignore perverse preferences of their constituents, but merely asks them not to voice them in the legislative chamber. The votes of representatives (if not their speeches) can still perfectly reflect constituent's desires.

That, however, may not be enough to satisfy constituents. Just as symbolic gestures (e.g. speeches on the floor of the House) are crucial components of the injury felt by those degraded by them, so too might they be crucial components in the satisfaction perversely derived by others in imposing such humiliations. In the 'rank order of discrimination' Myrdal (1944, pp. 60–1) discovered among racists in the American South, symbolic gestures are much more important than

---

[27] There is nothing undesirable about this hypocrisy, as Shklar (1979, p. 19) emphasizes: 'We assume that our public roles carry greater moral responsibilities than our private ones. We expect to behave better as citizens and public officials than as actors in the private sphere . . . It is, for example, no longer acceptable to make racist and anti-Semitic remarks in public in America; yet in private conversation racism and anti-Semitism are expressed as freely and as frequently as ever. Many a Southerner used to sneer at this display of hypocrisy. Now even he, like many a Northerner, is down to a few code words at election time. Would [anyone] prefer more public frankness? Should our public conduct really mirror our private, inner selves? Often our public manners are better than our personal laxities.'

material interests. Thus, perverse preferences might lead citizens to vote out representatives not only because they fail to vote their preferences, but also because they acquiesce in legislative procedures preventing them from voicing their preferences.

The political feasibility of any proposal for restricting majority rule must ultimately depend, directly or indirectly, upon the acquiescence of the majority itself. We are accustomed to speaking as if a 'scheme of civil rights' will necessarily 'remove those decisions from majoritarian political institutions altogether' (Dworkin 1978, p. 134). But, of course, there are no grounds for such confidence. Constitutions can be amended or replaced, and rights within them withdrawn. A politician who persists in playing by the established rules when his constituents insistently demand that he refrain from doing so, will simply be replaced (Barry 1975, p. 409; Goodin 1975). It is, therefore, a somewhat surprising and hopeful sign that the low popular regard for various American constitutional rights (Prothro and Grigg 1960; McClosky 1964) has translated into so few serious political demands for their repeal. This shows that majorities with undeniably perverse preferences might, under some circumstances, abide by (and perhaps even appreciate) institutions denying them the opportunity to put those preferences into practice.

Politically, the best explanation for why people are prepared to put up with such restrictions is probably couched in terms of the 'reciprocal forbearance' model (see Section II above). Having their own preferences denied—either by rights restricting outputs or by rules restricting inputs—is merely the price they pay for institutions that will similarly deny the preferences of others.[28] There is no more point in punishing their representatives for agreeing to these restrictions than there is in punishing them for agreeing to a national budget that devotes less than 100 per cent of public expenditures to their own

[28] This may be because people fear that they will need the protection of such guarantees themselves on later occasions. On this model, 'persistent minorities' would pose no threats and, hence, enjoy no protections. (The problem is not so much their minority status as the guaranteed persistence of the dimension of cleavage along which they constitute a minority.) Alternatively, people might agree to restrictions on their preferences because they foresee that the eventual consequences of respecting such preferences for everyone is unacceptable. Everyone pursuing 'status goods', for example, is self-defeating (Hirsch 1976), so people may agree to rule out such preferences from the outset. Similarly, racial hatred is self-fuelling (Lave and March 1975, pp. 396–402), so people might agree that racist preferences be ignored in order to nip the process in the bud.

constituency. People rightly regard these as constraints growing out of the nature of social life, for which no one deserves credit or blame, and which no one can repudiate without removing himself from the protection of civil society. Sometimes this is a group's goal—Irish Nationalist MPs flagrantly and persistently violated the rules of debate in the British House of Commons with the explicit intention of getting themselves expelled from the chamber and, ultimately, their province cast out of the Union (Taylor 1971, pp. 98–9). Beyond such exceptional cases, however, we can probably count on pragmatic acceptance of preference-laundering devices as well as vested rights guarantees.

V. Conclusion

Social choice theorists, in order to make their subject mathematically tractable, narrow 'utility' to a simple preference-ordering. Much is gained in terms of formal rigour. But much is also lost. For some purposes, a broader understanding seems far more useful. The application that has attracted my own attention is the laundering of preferences—both in explaining how individuals launder the preferences they express in collective decision contexts and in justifying social authorities in responding selectively to only certain sorts of citizen preferences. This is one way in which a broader understanding of 'utility information' might be useful. Doubtless there are many others.

REFERENCES

Anscombe, G. E. M. (1978) 'On the source of the authority of the state', *Ratio* 20, 1–28.

Arrow, K. J. (1963) *Social Choice and Individual Values*, 2nd edn, New Haven: Yale University Press.

Arrow, K. J. (1967) 'The place of moral obligation in preference systems', in S. Hook (ed.), *Human Values and Economic Policy*, New York: New York University Press, pp. 117–19.

Banfield, E. and Wilson, J. Q. (1964) 'Public regardingness as a value premise in voting behaviour', *American Political Science Review* 63, 876–87.

Barry, B. (1965) *Political Argument*, London: Routledge and Kegan Paul.

Barry, B. (1975) 'The consociational model and its dangers', *European Journal of Political Research* 3, 393–412.

Bator, F. M. (1957) 'The simple analytics of welfare maximization', *American Economic Review* 47, 22–59.

Benn, S. I. (1971) 'Privacy, freedom and respect for persons', in J. R. Pennock and J. W. Chapman (eds.), *Nomos XII: Privacy,* New York: Atherton, pp. 1–26.

Benn, S. (1976) 'Rationality and political behaviour', in G. W. Mortimore and S. I. Benn (eds.), *Rationality and the Social Sciences*, London: Routledge and Kegan Paul, pp. 246–7.

Benn, S. I. (1979) 'The problematic rationality of political participation', in P. Laslett and J. Fishkin (eds.), *Philosophy, Politics and Society*, 5th series, Oxford: Blackwell, pp. 291–312.

Bilmes, J. M. (1979) 'The evolution of decisions in a Thai village: a quasi-experimental study', *Human Organization* 38, 169–78.

Brandt, R. B. (1967) 'Personal values and the justification of institutions', in S. Hook (ed.), *Human Values and Economic Policy*, New York: New York University Press, pp. 22–40.

Broome, J. (1978) 'Choice and value in economics', *Oxford Economic Papers* 30, 313–33.

Cannon, C. (1963) *Procedure in the House of Representatives*, 87th Congress, 2nd Session, Washington, D.C.: Government Printing Office.

Downs, A. (1957) *An Economic Theory of Democracy*, New York: Harper.

Dunleavy, P. and Ward, H. (1981) 'Exogenous voter preferences and parties with state power', *British Journal of Political Science* 11, 351–80.

Dworkin, G. (1971) 'Paternalism', in R. A. Wasserstrom, (ed.), *Morality and the Law*, Belmont, Calif.: Wadsworth, pp. 107–26.

Dworkin, R. M. (1977) *Taking Rights Seriously*, London: Duckworth.

Dworkin, R. M. (1978) 'Liberalism', in S. Hampshire (ed.), *Public and Private Morality*, Cambridge: Cambridge University Press, pp. 113–43.

Dworkin, R. M. (1981) 'What is equality? Part 1: equality of welfare', *Philosophy and Public Affairs* 10, 185–246.

Eckstein, O. (1961) 'A survey of the theory of public expenditure criteria', in National Bureau of Economic Research, *Public Finances: Needs, Sources and Utilization*, Princeton: Princeton University Press, pp. 439–94.

Elster, J. (1976) 'Some conceptual problems in political theory', in B. Barry (ed.), *Power and Political Theory*, London: Wiley, pp. 245–70.

Fiorina, M. P. (1981) *Restrospective Voting in American National Elections*, New Haven: Yale University Press.

Frankfurt, H. G. (1971) 'Freedom of the will and the concept of a person', *Journal of Philosophy* 68, 5–20.

Fried, C. (1978) *Right and Wrong*, Cambridge Mass.: Harvard University Press.

Goodin, R. E. (1975) 'Cross-cutting cleavages and social conflict', *British Journal of Political Science* 5, 516–19.

Goodin, R. E. (1980) *Manipulatory Politics*, New Haven: Yale University Press.

Goodin, R. E. (1982) *Political Theory and Public Policy*, Chicago: University of Chicago Press.

Goodin, R. E. and Roberts, K. W. S. (1975) 'The ethical voter', *American Political Science Review* 69, 926–8.

Gray, J. (1981) 'John Stuart Mill on liberty, utility and rights', in J. R. Pennock and J. W. Chapman (eds.), *Nomos XXIII: Human Rights*, New York: New York University Press, pp. 80–116.

Gross, B. M. (1953) *The Legislative Struggle*, New York: McGraw–Hill.

Harsanyi, J. C. (1955) 'Cardinal welfare, individualistic ethics and interpersonal comparisons of utility', *Journal of Political Economy* 63, 309–21.

Harsanyi, J. C. (1977) *Rational Behavior and Bargaining Equilibrium in Games and Social Situations*, Cambridge: Cambridge University Press.

Hart, H. L. A. (1955) 'Are there any natural rights?' *Philosophical Review* 64, 175–91.

Hart, H. L. A. (1979) 'Between utility and rights', in A. Ryan (ed.), *The Idea of Freedom*, Oxford: Clarendon Press, pp. 77–98.

Hayek, F. A. (1960) *The Constitution of Liberty*, London: Routledge and Kegan Paul.

Head, J. G. (1974) *Public Goods and Public Welfare*, Durham, N.C.: Duke University Press.

Hill, T. E., Jr. (1979) 'Symbolic protest and calculated silence', *Philosophy and Public Affairs* 9, 83–102.

Hirsch, F. (1976) *Social Limits to Growth*, London: Routledge and Kegan Paul.

Hollis, M. (1977) *Models of Man*, Cambridge: Cambridge University Press.

Hollis, M. (1981) 'Economic man and original sin', *Political Studies* 29, 167–80.

Hume, D. (1739) *A Treatise of Human Nature*, London: John Noon.

Hume, D. (1777) *An Enquiry Concerning the Principles of Morals*, London: T. Cadell.

Jefferson, T. (1801) *Manual of Parliamentary Practice*, in G. F. Harrison and J. P. Coder (eds.), *Senate Manual*, 89th Congress, 1st Session, Washington, D.C.: Government Printing Office, 1965, pp. 700–54.

Kinder, D. R. and Kiewiet, D. R. (1981) 'Sociotropic politics: the American case', *British Journal of Political Science* 11, 129–61.

Lancaster, K. J. (1967) 'A new approach to consumer theory', *Journal of Political Economy* 75, 132–57.

Lave, C. A. and March, J. G. (1975) *An Introduction to Models in the Social Sciences*, New York: Harper and Row.

Little, I. M. D. (1952) 'Social choice and individual values', *Journal of Political Economy* 60, 422–32.

Maass, A. (1966) 'Benefit-cost analysis: its relevance to public investment decisions', *Quarterly Journal of Economics* 80, 208–26.

MacRae, D., Jr. (1976) *The Social Function of Social Science*, New Haven: Yale University Press.

Marglin, S. A. (1963) 'The social rate of discount and the optimal rate of investment', *Quarterly Journal of Economics* 77, 95–11.

Margolis, H. (1981) 'A new model of rational choice', *Ethics* 91, 265–79.

May, E. (1971) *The Law, Privileges, Proceedings and Usage of Parliament*, B. Cocks (ed.), 18th edn, London: Butterworth.

Mayo, H. B. (1960) *An Introduction to Democratic Theory*, New York: Oxford University Press.

McClosky, H. (1964) 'Consensus and ideology in American politics', *American Political Science Review* 58, 361–82.

Mill, J. S. (1859) *On Liberty*, London: Parker and Son.

Musgrave, R. A. (1968) 'Public expenditures', in D. L. Sills (ed.), *International Encyclopaedia of the Social Sciences*, London: Collier-Macmillan, Vol. 13, pp. 156–64.

Myrdal, G. (1944) *An American Dilemma*, London: Harper.

Parfit, D. (1976) 'Lewis, Perry and what matters', in A. O. Rorty (ed.), *The Identities of Persons*, Berkeley: University of California Press, pp. 91–108.

Pigou, A. C. (1932) *The Economics of Welfare*, 4th edn, London: Macmillan.

Plamenatz, J. (1973) *Democracy and Illusion*, London: Longman.

Prothro, J. W. and Grigg, C. M. (1960) 'Fundamental principles of democracy: bases of agreement and disagreement', *Journal of Politics* 22, 276–94.

Raz, J. (1975) *Practical Reasons and Norms*, London: Hutchinson.

Rothenberg, J. (1968) 'Consumer sovereignty', in D. L. Sills (ed.), *International Encyclopaedia of the Social Sciences*, London: Collier-Macmillan, Vol. 3, pp. 326–35.

Rousseau, J.-J. (1762) *On the Social Contract*, ed. and trans. R. D. and J. R. Masters, New York: St. Martin's Press, 1978.

Runciman, W. G. and Sen, A. K. (1965) 'Games, justice and the general will', *Mind* 74, 554–62.

Sears, D. O., Lau, R. R., Taylor, T. R. and Allen, H. M., Jr. (1980) 'Self-interest vs. symbolic politics in policy attitudes and presidential voting', *American Political Science Review* 74, 670–84.

Self, P. (1975) *Econocrats and the Policy Process*, London: Macmillan.

Sen, A. K. (1970) *Collective Choice and Social Welfare*, San Francisco: Holden–Day.

Sen, A. K. (1973) 'Behaviour and the concept of preference', *Economica* 40, 241–59.

Sen, A. K. (1974) 'Choice, ordering and morality', in S. Körner (ed.), *Practical Reason*, Oxford: Blackwell, pp. 54–67.

Sen, A. K. (1976) 'Liberty, unanimity and rights', *Economica* 43, 217–45.

Sen, A. K. (1977) 'Rational fools', *Philosophical and Public Affairs* 6, 317–44.

Sen, A. K. (1979a) 'Personal utilities and public judgments: or what's wrong with welfare economics?', *Economic Journal* 89, 537–58.

Sen, A. K. (1979b) 'Utilitarianism and welfarism', *Journal of Philosophy* 76, 463–89.

Sen, A. K. (1980–81) 'Plural utilities', *Proceedings of the Aristotelian Society* 81, 193–215.

Shklar, J. (1979) 'Let us not be hypocritical', *Daedalus* 108, 1–25.

Skinner, B. F. (1948) *Walden II*, New York: Macmillan.

Smart, J. J. C. (1973) 'An outline of a system of utilitarian ethics', in J. J. C. Smart and B. Williams, *Utilitarianism, For and Against*, Cambridge: Cambridge University Press, pp. 3–174.

Smith, A. (1790) *The Theory of Moral Sentiments*, 6th edn, London: A. Strahan and T. Cadell.

Stephen, J. F. (1874) *Liberty, Equality, Fraternity*, 2nd edn, London: Smith and Elder.

Taylor, E. (1971) *The House of Commons at Work*, 8th edn, Harmondsworth: Penguin.

Tversky, A. and Kahneman, D. (1981) 'The framing of decisions and the psychology of choice', *Science* 211, 453–8.

US Congress. Senate (1965) 'Standing rules for conducting business in the Senate of the United States', in G. F. Harrison and J. P. Coder (eds.), *Senate Manual*, 89th Congress, 1st Session, Washington, D.C.: Government Printing Office, pp. 1–41.

Williams, B. (1973) 'A critique of utilitarianism', in J. J. C. Smart and B. Williams, *Utilitarianism, For and Against*, Cambridge: Cambridge University Press, pp. 75–150.

Wollheim, R. (1958) 'Democracy', *Journal of the History of Ideas* 19, 225–42.

4. The market and the forum: three varieties of political theory

JON ELSTER

I want to compare three views of politics generally, and of the democratic system more specifically. I shall first look at social choice theory, as an instance of a wider class of theories with certain common features. In particular, they share the conception that the political process is instrumental rather than an end in itself, and the view that the decisive political act is a private rather than a public action, viz. the individual and secret vote. With these usually goes the idea that the goal of politics is the optimal compromise between given, and irreducibly opposed, private interests. The other two views arise when one denies, first, the private character of political behaviour and then, secondly, goes on also to deny the instrumental nature of politics. According to the theory of Jürgen Habermas, the goal of politics should be rational agreement rather than compromise, and the decisive political act is that of engaging in public debate with a view to the emergence of a consensus. According to the theorists of participatory democracy, from John Stuart Mill to Carole Pateman, the goal of politics is the transformation and education of the participants. Politics, on this view, is an end in itself—indeed many have argued that it represents the good life for man. I shall discuss these views in the order indicated. I shall present them in a somewhat stylized form, but my critical comments will not I hope, be directed to strawmen.

I

Politics, it is usually agreed, is concerned with the common good, and notably with the cases in which it cannot be realized as the aggregate outcome of individuals pursuing their private interests. In particular, uncoordinated private choices may lead to outcomes that are worse for all than some other outcome that could have been attained by coordination. Political institutions are set up to remedy such *market failures*, a phrase that can be taken either in the static sense of an inability to provide public goods or in the more dynamic sense of a breakdown of the self-regulating properties usually ascribed to the market mechanism.[1] In addition there is the redistributive task of politics—moving along the Pareto-optimal frontier once it has been reached.[2] According to the first view of politics, this task is inherently one of interest struggle and compromise. The obstacle to agreement is not only that most individuals want redistribution to be in their favour, or at least not in their disfavour.[3] More basically consensus is blocked because there is no reason to expect that individuals will converge in their views on what constitutes a just redistribution.

I shall consider social choice theory as representative of the private-instrumental view of politics, because it brings out supremely well the logic as well as the limits of that approach. Other varieties, such as the Schumpeterian or neo-Schumpeterian theories, are closer to the actual political process, but for that reason also less suited to my purpose. For instance, Schumpeter's insistence that voter preferences are shaped and manipulated by politicians[4] tends to blur the distinction, central to my analysis, between politics as the aggregation of given preferences and politics as the transformation of preferences through rational discussion. And although the neo-Schumpeterians are right in emphasizing the role of the political parties in the preference-aggregation process,[5] I am not here concerned with such mediating mechanisms. In any case, political problems also arise within the political

---

[1] Elster (1978, Ch. 5) refers to these two varieties of market failure as *suboptimality* and *counterfinality* respectively, linking them both to collective action.

[2] This is a simplification. First, as argued in Samuelson (1950), there may be political constraints that prevent one from attaining the Pareto-efficient frontier. Secondly, the very existence of several points that are Pareto-superior to the *status quo*, yet involve differential benefits to the participants, may block the realization of any of them.

[3] Hammond (1976) offers a useful analysis of the consequences of selfish preferences over income distributions, showing that 'without interpersonal comparisons of some kind, any social preference ordering over the space of possible income distributions must be dictatorial'.

[4] Schumpeter (1961, p. 263): 'the will of the people is the product and not the motive power of the political process'. One should not, however, conclude (as does Lively 1975, p. 38) that Schumpeter thereby abandons the market analogy, since on his view (Schumpeter 1939, p. 73) consumer preferences are no less manipulable (with some qualifications stated in Elster 1983a, Ch. 5).

[5] See in particular Downs (1957).

parties, and so my discussion may be taken to apply to such lower-level political processes. In fact, much of what I shall say makes better sense for politics on a rather small scale—within the firm, the organization or the local community—than for nationwide political systems.

In very broad outline, the structure of social choice theory is as follows.[6] (1) We begin with a *given* set of agents, so that the issue of a normative justification of political boundaries does not arise. (2) We assume that the agents confront a *given* set of alternatives, so that for instance the issue of agenda manipulation does not arise. (3) The agents are supposed to be endowed with preferences that are similarly *given* and not subject to change in the course of the political process. They are, moreover, assumed to be causally independent of the set of alternatives. (4) In the standard version, which is so far the only operational version of the theory, preferences are assumed to be purely ordinal, so that it is not possible for an individual to express the intensity of his preferences, nor for an outside observer to compare preference intensities across individuals. (5) The individual preferences are assumed to be defined over all pairs of individuals, i.e. to be complete, and to have the formal property of transitivity, so that preference for A over B and for B over C implies preference for A over C.

Given this setting, the task of social choice theory is to arrive at a social preference ordering of the alternatives. This might appear to require more than is needed: why not define the goal as one of arriving at the choice of one alternative? There is, however, usually some uncertainty as to which alternatives are really feasible, and so it is useful to have an ordering if the top-ranked alternative proves unavailable. The ordering should satisfy the following criteria. (6) Like the individual preferences, it should be complete and transitive. (7) It should be Pareto-optimal, in the sense of never having one option socially preferred to another which is individually preferred by everybody. (8) The social choice between two given options should depend only on how the individuals rank these two options, and thus not be sensitive to changes in their preferences concerning other options. (9) The social preference ordering should respect and reflect individual preferences, over and above the condition of Pareto-

---

[6] For fuller statements, see Arrow (1963), Sen (1970), and Kelly (1978), as well as the contribution of Aanund Hylland to the present volume.

optimality. This idea covers a variety of notions, the most important of which are *anonymity* (all individuals should count equally), *non-dictatorship* (*a fortiori* no single individual should dictate the social choice), *liberalism* (all individuals should have some private domain within which their preferences are decisive), and *strategy-proofness* (it should not pay to express false preferences).

The substance of social choice theory is given in a series of impossibility and uniqueness theorems, stating either that a given subset of these conditions is incapable of simultaneous satisfaction or that they uniquely describe a specific method for aggregating preferences. Much attention has been given to the impossibility theorems, yet from the present point of view these are not of decisive importance. They stem largely from the paucity of allowable information about the preferences, i.e. the exclusive focus on ordinal preferences.[7] True, at present we do not quite know how to go beyond ordinality. Log-rolling and vote-trading may capture some of the cardinal aspects of the preferences, but at some cost.[8] Yet even should the conceptual and technical obstacles to intra- and inter-individual comparison of preference intensity be overcome,[9] many objections to the social choice approach would remain. I shall discuss two sets of objections, both related to the assumption of given preferences. I shall argue, first, that the preferences people choose to express may not be a good guide to what they really prefer; and secondly that what they really prefer may in any case be a fragile foundation for social choice.

In actual fact, preferences are never 'given', in the sense of being directly observable. If they are to serve as inputs to the social choice process, they must somehow be *expressed* by the individuals. The expression of preferences is an action, which presumably is guided by these very same preferences.[10] It is then far from obvious that the individually rational action is to express these preferences as they are. Some methods for aggregating preferences are such that it may pay the individual to express false preferences, i.e. the outcome may in some cases be better according to his real preferences if he chooses not to

---

[7] Cf. d'Aspremont and Gevers (1977).

[8] Riker and Ordeshook (1973, pp. 112–13).

[9] Cf. the contributions of Donald Davidson and Allan Gibbard to the present volume.

[10] Presumably, but not obviously, since the agent might have several preference orderings and rely on higher-order preferences to determine which of the first-order preferences to express, as suggested for instance by Sen (1976).

express them truthfully. The condition for strategy-proofness for social choice mechanisms was designed expressly to exclude this possibility. It turns out, however, that the systems in which honesty always pays are rather unattractive in other respects.[11] We then have to face the possibility that even if we require that the social preferences be Pareto-optimal with respect to the expressed preferences, they might not be so with respect to the real ones. Strategy-proofness and collective rationality, therefore, stand and fall together. Since it appears that the first must fall, so must the second. It then becomes very difficult indeed to defend the idea that the outcome of the social choice mechanism represents the common good, since there is a chance that everybody might prefer some other outcome.

Amos Tversky has pointed to another reason why choices—or expressed preferences—cannot be assumed to represent the real preferences in all case.[12] According to his 'concealed preference hypothesis', choices often conceal rather than reveal underlying preferences. This is especially so in two sorts of cases. First, there are the cases of anticipated regret associated with a risky decision. Consider the following example (from Tversky):

On her twelfth birthday, Judy was offered a choice between spending the weekend with her aunt in the city (C), or having a party for all her friends. The party could take place either in the garden (GP) or inside the house (HP). A garden party would be much more enjoyable, but there is always the possibility of rain, in which case an inside party would be more sensible. In evaluating the consequences of the three options, Judy notes that the weather condition does not have a significant effect on C. If she chooses the party, however, the situation is different. A garden party will be a lot of fun if the weather is good, but quite disastrous if it rains, in which case an inside party will be acceptable. The trouble is that Judy expects to have a lot of regret if the party is to be held inside and the weather is very nice.

[11] Pattanaik (1978) offers a survey of the known results. The only strategy-proof mechanisms for social choice turn out to be the dictatorial one (the dictator has no incentive to misrepresent his preferences) and the randomizing one of getting the probability that a given option will be chosen equal to the proportion of voters that have it as their first choice.
[12] Tversky (1981).

Now, let us suppose that for some reason it is no longer possible to have an outside party. In this situation, there is no longer any regret associated with holding an inside party in good weather because (in this case) Judy has no other place for holding the party. Hence, the elimination of an available course of action (holding the party outside) removes the regret associated with an inside party, and increases its overall utility. It stands to reason, in this case, that if Judy was indifferent between C and HP, in the presence of GP, she will prefer HP to C when GP is eliminated.

What we observe here is the violation of condition (8) above, the so-called 'independence of irrelevant alternatives'. The expressed preferences depend causally on the set of alternatives. We may assume that the real preferences, defined over the set of possible outcomes, remain constant, contrary to the case to be discussed below. Yet the preferences over the *pairs* (choice, outcome) depend on the set of available choices, because the 'costs of responsibility' differentially associated with various such pairs depend on what else one 'could have done'. Although Judy could not have escaped her predicament by deliberately making it physically impossible to have an outside party,[13] she might well have welcomed an event outside her control with the same consequence.

The second class of cases in which Tversky would want to distinguish the expressed preferences from the real preferences concerns decisions that are unpleasant rather than risky. For instance, 'society may prefer to save the life of one person rather than another, and yet be unable to make this choice'. In fact, losing both lives through inaction may be preferred to losing only one life by deliberate action. Such examples are closely related to the problems involved in act utilitarianism versus outcome utilitarianism.[14] One may well judge that it would be a good thing if state A came about, and yet not want to be the person by whose agency it comes about. The reasons for not wanting to be that person may be quite respectable, or they may not. The latter would be the case if one were afraid of being blamed by the relatives of the

---

[13] Cf. Elster (1979, Ch. II) or Schelling (1980) for the idea of deliberately restricting one's feasible set to make certain undesirable behaviour impossible at a later time. The reason this does not work here is that the regret would not be eliminated.

[14] Cf. for instance Williams (1973) or Sen (1979).

person who was deliberately allowed to die, or if one simply confused the causal and the moral notions of responsibility. In such cases the expressed preferences might lead to a choice that in a clear sense goes against the real preferences of the people concerned.

A second, perhaps more basic, difficulty is that the real preferences themselves might well depend causally on the feasible set. One instance is graphically provided by the fable of the fox and the sour grapes.[15] For the 'ordinal utilitarian', as Arrow for instance calls himself,[16] there would be no welfare loss if the fox were excluded from consumption of the grapes, since he thought them sour anyway. But of course the cause of his holding them to be sour was his conviction that he would in any case be excluded from consuming them, and then it is difficult to justify the allocation by invoking his preferences. Conversely, the phenomenon of 'counter-adaptive preferences'—the grass is always greener on the other side of the fence, and the forbidden fruit always sweeter—is also baffling for the social choice theorist, since it implies that such preferences, if respected, would not be satisfied—and yet the whole point of respecting them would be to give them a chance of satisfaction.

Adaptive and counter-adaptive preferences are only special cases of a more general class of desires, those which fail to satisfy some substantive criterion for acceptable preferences, as opposed to the purely formal criterion of transitivity. I shall discuss these under two headings: autonomy and morality.

Autonomy characterizes the way in which preferences are shaped rather than their actual content. Unfortunately I find myself unable to give a positive characterization of autonomous preferences, so I shall have to rely on two indirect approaches. First, autonomy is for desires what judgment is for belief. The notion of judgment is also difficult to define formally, but at least we know that there are persons who have this quality to a higher degree than others: people who are able to take account of vast and diffuse evidence that more or less clearly bears on the problem at hand, in such a way that no element is given undue importance. In such people the process of belief formation is not disturbed by defective cognitive processing, nor distorted by wishful thinking and the like. Similarly, autonomous preferences are those

---

[15] Cf. Elster (1983b, Ch. III) for a discussion of this notion.
[16] Arrow (1973).

that have not been shaped by irrelevant causal processes—a singularly unhelpful explanation. To improve somewhat on it, consider, secondly, a short list of such irrelevant causal processes. They include adaptive and counter-adaptive preferences, conformity and anti-conformity, the obsession with novelty and the equally unreasonable resistance to novelty. In other words, preferences may be shaped by adaptation to what is possible, to what other people do or to what one has been doing in the past—or they may be shaped by the desire to differ as much as possible from these. In all of these cases the source of preference change is not in the person, but outside him—detracting from his autonomy.

Morality, it goes without saying, is if anything even more controversial. (Within the Kantian tradition it would also be questioned whether it can be distinguished at all from autonomy.) Preferences are moral or immoral by virtue of their content, not by virtue of the way in which they have been shaped. Fairly uncontroversial examples of unethical preferences are spiteful and sadistic desires, and arguably also the desire for positional goods, i.e. goods such that it is logically impossible for more than a few to possess them.[17] The desire for an income twice the average can lead to less welfare for everybody, so that such preferences fail to pass the Kantian generalization test.[18] Also they are closely linked to spite, since one way of getting more than others is to take care that they get less—indeed this may often be a more efficient method than trying to excel.[19]

To see how the lack of autonomy may be distinguished from the lack of moral worth, let me use *conformity* as a technical term for a desire caused by a drive to be like other people, and *conformism* for a desire to be like other people, with anti-conformity and anti-conformism similarly defined. Conformity implies that other people's desires enter into the causation of my own, conformism that they enter irreducibly into the description of the object of my desires. Conformity may bring about conformism, but it may also lead to anti-conformism, as in

[17] Hirsch (1976).

[18] Haavelmo (1970) offers a model in which everybody may suffer a loss of welfare by trying to keep up with the neighbours.

[19] One may take the achievements of others as a parameter and one's own as the control variable, or conversely try to manipulate the achievements of others so that they fall short of one's own. The first of these ways of realizing positional goods is clearly less objectionable than the second, but still less pure than the non-comparative desire for a certain standard of excellence.

Theodore Zeldin's comment that among the French peasantry 'prestige is to a great extent obtained from conformity with traditions (so that the son of a non-conformist might be expected to be one too'.[20] Clearly, conformity may bring about desires that are morally laudable, yet lacking in autonomy. Conversely, I do not see how one could rule out on *a priori* grounds the possibility of autonomous spite, although I would welcome a proof that autonomy is incompatible not only with anti-conformity, but also with anti-conformism.

We can now state the objection to the political view underlying social choice theory. It is, basically, that it embodies a confusion between the kind of behaviour that is appropriate in the market place and that which is appropriate in the forum. The notion of consumer sovereignty is acceptable because, and to the extent that, the consumer chooses between courses of action that differ only in the way they affect him. In political choice situations, however, the citizen is asked to express his preference over states that also differ in the way in which they affect other people. This means that there is no similar justification for the corresponding notion of the citizen's sovereignty, since other people may legitimately object to social choice governed by preferences that are defective in some of the ways I have mentioned. A social choice mechanism is capable of resolving the market failures that would result from unbridled consumer sovereignty, but as a way of redistributing welfare it is hopelessly inadequate. If people affected each other only by tripping over each other's feet, or by dumping their garbage into one another's backyards, a social choice mechanism might cope. But the task of politics is not only to eliminate inefficiency, but also to create justice—a goal to which the aggregation of pre-political preferences is a quite incongruous means.

This suggests that the principles of the forum must differ from those of the market. A long-standing tradition from the Greek *polis* onwards suggests that politics must be an open and public activity, as distinct from the isolated and private expression of preferences that occurs in buying and selling. In the following sections I look at two different conceptions of public politics, increasingly removed from the market theory of politics. Before I go on to this, however, I should briefly consider an objection that the social choice theorist might well make to what has just been said. He could argue that the only alternative to the

[20] Zeldin (1973, p. 134).

aggregation of given preferences is some kind of censorship or paternalism. He might agree that spiteful and adaptive preferences are undesirable, but he would add that any institutional mechanism for eliminating them would be misused and harnessed to the private purposes of power-seeking individuals. Any remedy, in fact, would be worse than the disease. This objection assumes (i) that the only alternative to aggregation of given preferences is censorship, and (ii) that censorship is always objectionable. Robert Goodin, in his contribution to this volume, challenges the second assumption, by arguing that laundering or filtering of preferences by self-censorship is an acceptable alternative to aggregation. I shall now discuss a challenge to the first assumption, viz. the idea of a *transformation* of preferences through public and rational discussion.

II

Today this view is especially associated with the writings of Jürgen Habermas on 'the ethics of discourse' and 'the ideal speech situation'. As mentioned above, I shall present a somewhat stylized version of his views, although I hope they bear some resemblance to the original.[21] The core of the theory, then, is that rather than aggregating or filtering preferences, the political system should be set up with a view to changing them by public debate and confrontation. The input to the social choice mechanism would then not be the raw, quite possibly selfish or irrational, preferences that operate in the market, but informed and other-regarding preferences. Or rather, there would not be any need for an aggregating mechanism, since a rational discussion would tend to produce unanimous preferences. When the private and idiosyncratic wants have been shaped and purged in public discussion about the public good, uniquely determined rational desires would emerge. Not optimal compromise, but unanimous agreement is the goal of politics on this view.

There appear to be two main premises underlying this theory. The first is that there are certain arguments that simply cannot be stated publicly. In a political debate it is pragmatically impossible to argue

---

[21] I rely mainly on Habermas (1982). I also thank Helge Høibraaten, Rune Slagstad, and Gunnar Skirbekk for having patiently explained to me various aspects of Habermas's work.

that a given solution should be chosen just because it is good for oneself. By the very act of engaging in a public debate—by arguing rather than bargaining—one has ruled out the possibility of invoking such reasons.[22] To engage in discussion can in fact be seen as one kind of self-censorship, a pre-commitment to the idea of rational decision. Now, it might well be thought that this conclusion is too strong. The first argument only shows that in public debate one has to pay some lip-service to the common good. An additional premise states that over time one will in fact come to be swayed by considerations about the common good. One cannot indefinitely praise the common good 'du bout des lèvres', for—as argued by Pascal in the context of the wager—one will end up having the preferences that initially one was faking.[23] This is a psychological, not a conceptual premise. To explain why going through the motions of rational discussion should tend to bring about the real thing, one might argue that people tend to bring what they mean into line with what they say in order to reduce dissonance, but this is a dangerous argument to employ in the present context. Dissonance reduction does not tend to generate autonomous preferences. Rather one would have to invoke the power of reason to break down prejudice and selfishness. By speaking with the voice of reason, one is also exposing oneself to reason.

To sum up, the conceptual impossibility of expressing selfish arguments in a debate about the public good, and the psychological difficulty of expressing other-regarding preferences without ultimately coming to acquire them, jointly bring it about that public discussion tends to promote the common good. The *volonté générale*, then, will not simply be the Pareto-optimal realization of given (or expressed) preferences,[24] but the outcome of preferences that are themselves shaped by a concern for the common good. For instance, by mere aggregation of given preferences one would be able to take account of some negative externalities, but not of those affecting future generations. A social choice mechanism might prevent persons now living from dumping their garbage into one another's backyards, but not from dumping it on the future. Moreover, considerations of distributive justice within the Pareto constraint would now have a more solid

---

[22] Midgaard (1980).
[23] For Pascal's argument, cf. Elster (1979, Ch. II. 3).
[24] As suggested by Runciman and Sen (1965).

foundation, especially as one would also be able to avoid the problem of strategy-proofness. By one stroke one would achieve more rational preferences, as well as the guarantee that they will in fact be expressed.

I now want to set out a series of objections—seven altogether—to the view stated above. I should explain that the goal of this criticism is not to demolish the theory, but to locate some points that need to be fortified. I am, in fact, largely in sympathy with the fundamental tenets of the view, yet fear that it might be dismissed as Utopian, both in the sense of ignoring the problem of getting from here to there, and in the sense of neglecting some elementary facts of human psychology.

The *first objection* involves a reconsideration of the issues of paternalism. Would it not, in fact, be unwarranted interference to impose on the citizens the obligation to participate in political discussion? One might answer that there is a link between the right to vote and the obligation to participate in discussion, just as rights and duties are correlative in other cases. To acquire the right to vote, one has to perform certain civic duties that go beyond pushing the voting button on the television set. There would appear to be two different ideas underlying this answer. First, only those should have the right to vote who are sufficiently *concerned* about politics to be willing to devote some of their resources—time in particular—to it. Secondly, one should try to favour *informed* preferences as inputs to the voting process. The first argument favours participation and discussion as a sign of interest, but does not give it an instrumental value in itself. It would do just as well, for the purpose of this argument, to demand that people should pay for the right to vote. The second argument favours discussion as a means to improvement—it will not only select the right people, but actually make them more qualified to participate.

These arguments might have some validity in a near-ideal world, in which the concern for politics was evenly distributed across all relevant dimensions, but in the context of contemporary politics they miss the point. The people who survive a high threshold for participation are disproportionately found in a privileged part of the population. At best this could lead to paternalism, at worst the high ideals of rational discussion could create a self-elected elite whose members spend time on politics because they want power, not out of concern for the issues. As in other cases, to be discussed later, the best can be the enemy of the good. I am not saying that it is impossible to modify the ideal in a way that allows both for rational discussion and for low-profile participation, only that any institutional design must respect the trade-off between the two.

My *second objection* is that even assuming unlimited time for discussion, unanimous and rational agreement might not necessarily ensue. Could there not be legitimate and unresolvable differences of opinions over the nature of the common good? Could there not even be a plurality of ultimate values?

I am not going to discuss this objection, since it is in any case pre-empted by the *third objection*. Since there are in fact always time constraints on discussions—often the stronger the more important the issues—unanimity will rarely emerge. For any constellation of preferences short of unanimity, however, one would need a social choice mechanism to aggregate them. One can discuss only for so long, and then one has to make a decision, even if strong differences of opinion should remain. This objection, then, goes to show that the transformation of preferences can never do more than supplement the aggregation of preferences, never replace it altogether.

This much would no doubt be granted by most proponents of the theory. True, they would say, even if the ideal speech situation can never be fully realized, it will nevertheless improve the outcome of the political process if one goes some way towards it. The *fourth objection* questions the validity of this reply. In some cases a little discussion can be a dangerous thing, worse in fact than no discussion at all, viz. if it makes some but not all persons align themselves on the common good. The following story provides an illustration:

> Once upon a time two boys found a cake. One of them said, 'Splendid! I will eat the cake.' The other one said, 'No, that is not fair! We found the cake together, and we should share and share alike, half for you and half for me.' The first boy said, 'No, I should have the whole cake!' Along came an adult who said, 'Gentlemen, you shouldn't fight about this: you should *compromise*. Give him three quarters of the cake.'[25]

What creates the difficulty here is that the first boy's preferences are allowed to count twice in the social choice mechanism suggested by the

[25] Smullyan (1980, p. 56).

adult: once in his expression of them and then again in the other boy's internalized ethic of sharing. And one can argue that the outcome is socially inferior to that which would have emerged had they both stuck to their selfish preferences. When Adam Smith wrote that he had never known much good done by those who affected to trade for the public good, he may only have had in mind the harm that can be done by *unilateral* attempts to act morally. The categorical imperative itself may be badly served by people acting unilaterally on it.[26] Also, an inferior outcome may result if discussion brings about partial adherence to morality in all participants rather than full adherence in some and none in others, as in the story of the two boys. Thus Serge Kolm argues that economies with moderately altrustic agents tend to work less well than economies where either everybody is selfish or everybody is altruistic.[27]

A *fifth objection* is to question the implicit assumption that the body politic as a whole is better or wiser than the sum of its parts. Could it not rather be the case that people are made more, not less, selfish and irrational by interacting politically? The cognitive analogy suggests that the rationality of beliefs may be positively as well as negatively affected by interaction. On the one hand there is what Irving Janis has called 'group-think', i.e. mutually reinforcing bias.[28] On the other hand there certainly are many ways in which people can, and do, pool their opinions and supplement each other to arrive at a better estimate.[29] Similarly autonomy and morality could be enhanced as well as undermined by interaction. Against the pessimistic view of Reinhold Niebuhr that individuals in a group show more unrestrained egoism than in their personal relationships,[30] we may set Hannah Arendt's optimistic view:

American faith was not all based on a semireligious faith in human nature, but on the contrary, on the possibility of checking human nature in its singularity, by virtue of human bonds and mutual promises. The hope for man in his singularity lay in the fact that not man but men inhabit the earth and form a world between them. It is

[26] Sobel (1967).
[27] Kolm (1981a, b).
[28] Janis (1972).
[29] Cf. Hogarth (1977) and Lehrer (1978).
[30] Niebuhr (1932, p. 11).

human worldliness that will save men from the pitfalls of human nature.[31]

Niebuhr's argument suggests an aristocratic disdain of the *mass*, which transforms individually decent people—to use a characteristically condescending phrase—into an unthinking horde. While rejecting this as a general view, one should equally avoid the other extreme, suggested by Arendt. Neither the Greek nor the American assemblies were the paradigms of discursive reason that she makes them out to be. The Greeks were well aware that they might be tempted by demagogues, and in fact took extensive precautions against this tendency.[32] The American town surely has not always been the incarnation of collective freedom, since on occasion it could also serve as the springboard for witch hunts. The mere decision to engage in rational discussion does not ensure that the transactions will in fact be conducted rationally, since much depends on the structure and the framework of the proceedings. The random errors of selfish and private preferences may to some extent cancel each other out and thus be less to be feared than the massive and coordinated errors that may arise through group-think. On the other hand, it would be excessively stupid to rely on mutually compensating vices to bring about public benefits as a general rule. I am not arguing against the need for public discussion, only for the need to take the question of institutional and constitutional design very seriously.

A *sixth objection* is that unanimity, were it to be realized, might easily be due to conformity rather than to rational agreement. I would in fact tend to have more confidence in the outcome of a democratic decision if there was a minority that voted against it, than if it was unanimous. I am not here referring to people expressing the majority preferences against their real ones, since I am assuming that something like the secret ballot would prevent this. I have in mind that people may come to change their real preferences, as a result of seeing which way the majority goes. Social psychology has amply shown the strength of this bandwagon effect,[33] which in political theory is also known as the 'chameleon' problem.[34] It will not do to argue that the

---

[31] Arendt (1973, p. 174).
[32] Finley (1973); see also Elster (1979, Ch. II.8).
[33] Asch (1956) is a classic study.
[34] See Goldman (1972) for discussion and further references.

majority to which the conformist adapts his view is likely to pass the test of rationality even if his adherence to it does not, since the majority could well be made up of conformists each of whom would have broken out had there been a minority he could have espoused.

To bring the point home, consider a parallel case of non-autonomous preference formation. We are tempted to say that a man is free if he can get or do whatever it is that he wants to get or do. But then we are immediately faced with the objection that perhaps he only wants what he can get, as the result of some such mechanism as 'sour grapes'.[35] We may then add that, other things being equal, the person is freer the more things he wants to do which he is not free to do, since these show that his wants are not in general shaped by adaptation to his possibilities. Clearly, there is an air of paradox over the statement that a man's freedom is greater the more of his desires he is not free to realize, but on reflection the paradox embodies a valid argument. Similarly, it is possible to dissolve the air of paradox attached to the view that a collective decision is more trustworthy if it is less than unanimous.

My *seventh objection* amounts to a denial of the view that the need to couch one's argument in terms of the common good will purge the desires of all selfish arguments. There are in general many ways of realizing the common good, if by that phrase we now only mean some arrangement that is Pareto-superior to uncoordinated individual decisions. Each such arrangement will, in addition to promoting the general interest, bring an extra premium to some specific group, which will then have a strong interest in that particular arrangement.[36] The group may then come to prefer the arrangement because of that premium, although it will argue for it in terms of the common good. Typically the arrangement will be justified by a causal theory—an account, say, of how the economy works—that shows it to be not only *a* way, but the only way of promoting the common good. The economic theories underlying the early Reagan administration provide an example. I am not imputing insincerity to the proponents of these views, but there may well be an element of wishful thinking. Since social scientists disagree so strongly among themselves as to how societies work, what could be more human than to pick on a theory that uniquely justifies

---

[35] Berlin (1969, p. xxxviii); cf. also Elster (1983b, Ch. III.3).

[36] Schotter (1981, pp. 26 ff., pp. 43 ff.) has a good discussion of this predicament.

the arrangement from which one stands to profit? The opposition between general interest and special interests is too simplistic, since the private benefits may causally determine the way in which one conceives of the common good.

These objections have been concerned to bring out two main ideas. First, one cannot assume that one will in fact approach the good society by acting as if one had already arrived there. The fallacy inherent in this 'approximation assumption'[37] was exposed a long time ago in the economic 'theory of the second best':

> It is *not* true that a situation in which more, but not all, of the optimum conditions are fulfilled is necessarily, or is even likely to be, superior to a situation in which fewer are fulfilled. It follows, therefore, that in a situation in which there exist many constraints which prevent the fulfilment of the Paretian optimum conditions, the removal of any one constraint may affect welfare or efficiency either by raising it, by lowering it or by leaving it unchanged.[38]

The ethical analogue is not the familiar idea that some moral obligations may be suspended when other people act non-morally.[39] Rather it is that the nature of the moral obligation is changed in a non-moral environment. When others act non-morally, there may be an obligation to deviate not only from what they do, but also from the behaviour that would have been optimal if adopted by everybody.[40] In particular, a little discussion, like a little rationality or a little socialism, may be a dangerous thing.[41] If, as suggested by Habermas, free and rational discussion will only be possible in a society that has abolished political and economic domination, it is by no means obvious that abolition can be brought about by rational argumentation. I do not want to suggest that it could occur by force—since the use of force to end the use of

---

[37] Margalit (1983).
[38] Lipsey and Lancaster (1956–7, p. 12).
[39] This is the point emphasized in Lyons (1965).
[40] Cf. Hansson (1970) as well as Føllesdal and Hilpinen (1971) for discussions of 'conditional obligations' within the framework of deontic logic. It does not appear, however, that the framework can easily accommodate the kind of dilemma I am concerned with here.
[41] Cf. for instance Kolm (1977) concerning the dangers of a piecemeal introduction of socialism—also mentioned by Margalit (1983) as an objection to Popper's strategy for piecemeal social engineering.

force is open to obvious objections. Yet something like irony, eloquence or propaganda might be needed, involving less respect for the interlocutor than what would prevail in the ideal speech situation.

As will be clear from these remarks, there is a strong tension between two ways of looking at the relation between political ends and means. On the one hand, the means should partake of the nature of the ends, since otherwise the use of unsuitable means might tend to corrupt the end. On the other hand, there are dangers involved in choosing means immediately derived from the goal to be realized, since in a non-ideal situation these might take us away from the end rather than towards it. A delicate balance will have to be struck between these two, opposing considerations. It is in fact an open question whether there exists a ridge along which we can move to the good society, and if so whether it is like a knife-edge or more like a plateau.

The second general idea that emerges from the discussion is that even in the good society, should we hit upon it, the process of rational discussion could be fragile, and vulnerable to adaptive preferences, conformity, wishful thinking and the like. To ensure stability and robustness there is a need for structures—political institutions or constitutions—that could easily reintroduce an element of domination. We would in fact be confronted, at the political level, with a perennial dilemma of individual behaviour. How is it possible to ensure at the same time that one is bound by rules that protect one from irrational or unethical behaviour—and that these rules do not turn into prisons from which it is not possible to break out even when it would be rational to do so?[42]

III

It is clear from Habermas's theory, I believe, that rational political discussion has an *object* in terms of which it makes sense.[43] Politics is concerned with substantive decision-making, and is to that extent instrumental. True, the idea of instrumental politics might also be taken in a more narrow sense, as implying that the political process is

---

[42] Cf. Ainslie (1982) and Elster (1979, Ch. II.9).

[43] Indeed, Habermas (1982) is largely concerned with maxims for *action*, not with the evaluation of states of affairs.

one in which individuals pursue their selfish interests, but more broadly understood it implies only that political action is primarily a means to a non-political end, only secondarily, if at all, an end in itself. In this section I shall consider theories that suggest a reversal of this priority, and that find the main point of politics in the educative or otherwise beneficial effects on the participants. And I shall try to show that this view tends to be internally incoherent, or self-defeating. The benefits of participation are by-products of political activity. Moreover, they are *essentially* by-products, in the sense that any attempt to turn them into the main purpose of such activity would make them evaporate.[44] It can indeed be highly satisfactory to engage in political work, but only on the condition that the work is defined by a serious purpose which goes beyond that of achieving this satisfaction. If that condition is not fulfilled, we get a narcissistic view of politics— corresponding to various consciousness-raising activities familiar from the last decade or so.

My concern, however, is with political theory rather than with political activism. I shall argue that certain types of arguments for political institutions and constitutions are self-defeating, since they justify the arrangement in question by effects that are essentially by-products. Here an initial and important distinction must be drawn between the task of justifying a constitution *ex ante* and that of evaluating it *ex post* and at a distance. I argue below that Tocqueville, when assessing the American democracy, praised it for consequences that are indeed by-products. In his case, this made perfectly good sense as an analytical attitude adopted after the fact and at some distance from the system he was examining. The incoherence arises when one invokes the same arguments before the fact, in public discussion. Although the constitution-makers may secretly have such side effects in mind, they cannot coherently invoke them in public.

Kant proposed a *transcendental formula of public right*: 'All actions affecting the rights of other human beings are wrong if their maxim is not compatible with their being made public.'[45] Since Kant's illustrations of the principle are obscure, let me turn instead to John Rawls, who imposes a similar condition of publicity as a constraint on what the

---

[44] Cf. Elster (1983b, Ch. III) for a discussion of the notion that some psychological or social states are essentially by-products of actions undertaken for some other purpose.
[45] Kant (1795, p. 126).

parties can choose in the original position.[46] He argues, moreover, that this condition tends to favour his own conception of justice, as compared to that of the utilitarians.[47] If utilitarian principles of justice were openly adopted, they would entail some loss of self-esteem, since people would feel that they were not fully being treated as ends in themselves. Other things being equal, this would also lead to a loss in average utility. It is then conceivable that public adoption of Rawls's two principles of justice would bring about a higher average utility than public adoption of utilitarianism, although a lower average than under a secret utilitarian constitution introduced from above. The latter possibility, however, is ruled out by the publicity constraint. A utilitarian could not then advocate Rawls's two principles on utilitarian grounds, although he might well applaud them on such grounds. The fact that the two principles maximize utility would essentially be a by-product, and if chosen on the grounds that they are utility-maximizing they would no longer be so. Utilitarianism, therefore, is self-defeating in Kant's sense: 'it essentially lacks openness'.[48]

Derek Parfit has raised a similar objection to act consequentialism (AC) and suggested how it could be met:

> This gives to all one common aim: the best possible outcome. If we try to achieve this, we may often fail. Even when we succeed, the fact that we are disposed to try might make the outcome worse. AC might thus be indirectly self-defeating. What does this show? A consequentialist might say: 'It shows that AC should be only one part of our moral theory. It should be the part that covers successful acts. When we are certain to succeed, we should aim for the best possible outcome. Our wider theory should be this: we should have the aim and dispositions having which would make the outcome best. This wider theory would not be self-defeating. So the objection has been met.'[49]

Yet there is an ambiguity in the word 'should' in the penultimate sentence, since it is not clear whether we are told that it is good to have certain aims and dispositions, or that we should aim at having them. The latter answer immediately raises the problem that having certain

---

[46] Rawls (1971, p. 133).   [47] Rawls (1971, pp. 177 ff., esp. p. 181).
[48] Williams (1973, p. 123).   [49] Parfit (1981, p. 554).

aims and dispositions—i.e. being a certain kind of person—is essentially a by-product. When instrumental rationality is self-defeating, we cannot decide on instrumentalist grounds to take leave of it—no more than we can fall asleep by deciding not to try to fall asleep. Although spontaneity may be highly valuable on utilitarian grounds, 'you cannot both genuinely possess this kind of quality and also reassure yourself that while it is free and creative and uncalculative, it is also acting for the best'.[50]

Tocqueville, in a seeming paradox, suggested that democracies are less suited than aristocracies to deal with long-term planning, and yet are superior in the long-run to the latter. The paradox dissolves once it is seen that the first statement involves time at the level of the actors, the second at the level of the observer. On the one hand, 'a democracy finds it difficult to coordinate the details of a great undertaking and to fix on some plan and carry it through with determination in spite of obstacles. It has little capacity for combining measures in secret and waiting patiently for the result'.[51] On the other hand, 'in the long run government by democracy should increase the real forces of a society, but it cannot immediately assemble at one point and at a given time, forces as great as those at the disposal of an aristocratic government'.[52] The latter view is further elaborated in a passage from the chapter on 'The Real Advantages Derived by American Society from Democratic Government':

That constantly renewed agitation introduced by democratic government into political life passes, then, into civil society. Perhaps, taking everything into consideration, that is the greatest advantage of democratic government, and I praise it much more on account of what it causes to be done than for what it does. It is incontestable that the people often manage public affairs very badly, but their concern therewith is bound to extend their mental horizon and to shake them out of the rut of ordinary routine . . . Democracy does not provide a people with the most skillful of governments, but it does that which the most skillful government often cannot do: it spreads throughout the body social a restless activity, superabundant force, and energy never found elsewhere, which, however

---

[50] Williams (1973, p. 131); also Elster (1983b, Ch. II.3).
[51] Tocqueville (1969, p. 229).   [52] Tocqueville (1969, p. 224).

little favoured by circumstances, can do wonders. Those are its true advantages.[53]

The advantages of democracies, in other words, are mainly and essentially by-products. The avowed aim of democracy is to be a good system of government, but Tocqueville argues that it is inferior in this respect to aristocracy, viewed purely as a decision-making apparatus. Yet the very activity of governing democratically has as a by-product a certain energy and restlessness that benefits industry and generates prosperity. Assuming the soundness of this observation, could it ever serve as a public justification for introducing democracy in a nation that had not yet acquired it? The question is somewhat more complex than one might be led to think from what I have said so far, since the quality of the decisions is not the only consideration that is relevant for the choice of a political system. The argument from *justice* could also be decisive. Yet the following conclusion seems inescapable: if the system has no inherent advantage in terms of justice or efficiency, one cannot coherently and publicly advocate its introduction because of the side effects that would follow in its wake. There must be a *point* in democracy as such. If people are motivated by such inherent advantages to throw themselves into the system, other benefits may ensue—but the latter cannot by themselves be the motivating force. If the democratic method is introduced in a society solely because of the side effects on economic prosperity, and no one believes in it on any other ground, it will not produce them.

Tocqueville, however, did not argue that political activity is an end in itself. The justification for democracy is found in its effects, although not in the intended ones, as the strictly instrumental view would have it. More to the point is Tocqueville's argument for the jury system: 'I do not know whether a jury is useful to the litigants, but I am sure that it is very good for those who have to decide the case. I regard it as one of the most effective means of popular education at society's disposal.'[54] This is still an instrumental view, but the gap between the means and the end is smaller. Tocqueville never argued that the effect of democracy was to make politicians prosperous, only that it was conducive to general prosperity. By contrast, the justification of the jury system is found in the effect on the jurors themselves. And, as

[53] Tocqueville (1969, pp. 243–4).   [54] Tocqueville (1969, p. 275).

above, that effect would be spoilt if they believed that the impact on their own civic spirit was the main point of the proceedings.

John Stuart Mill not only applauded but advocated democracy on the ground of such educative effects on the participants. In current discussion he stands out both as an opponent of the purely instrumental view of politics, that of his father James Mill,[55] and as a forerunner of the theory of participatory democracy.[56] In his theory the gap between means and ends in politics is even narrower, since he saw political activity not only as a means to self-improvement, but also as a source of satisfaction and thus a good in itself. As noted by Albert Hirschman, this implies that 'the benefit of collective action for an individual is not the difference between the hoped-for result and the effort furnished by him or her, but the *sum* of these two magnitudes'.[57] Yet this very way of paraphrasing Mill's view also points to a difficulty. Could it really be the case that participation would yield a benefit even when the hoped-for results are nil, as suggested by Hirschman's formula? Is it not rather true that the effort is itself a function of the hoped-for result, so that in the end the latter is the only independent variable? When Mill refers, critically, to the limitations of Bentham, whose philosophy 'can teach the means of organising and regulating the merely *business* part of the social arrangement',[58] he seems to be putting the cart before the horse. The non-business part of politics may be the more valuable, but the value is contingent on the importance of the business part.

For a fully developed version of the non-instrumental theory of politics, we may go to the work of Hannah Arendt. Writing about the distinction between the private and the public realm in ancient Greece, she argues that:

Without mastering the necessities of life in the household, neither life nor the 'good life' is possible, but politics is never for the sake of life. As far as the members of the *polis* are concerned, household life exists for the sake of the 'good life' in the *polis*.[59]

[55] Cf. Ryan (1972). His contrast between 'two concepts of democracy' corresponds in part to the distinction between the first and the second of the theories discussed here, in part to the distinction between the first and the third, as he does not clearly separate the public conception of politics from the non-instrumental one.
[56] Pateman (1970, p. 29).   [57] Hirschman (1982, p. 82).
[58] Mill (1859, p. 105).   [59] Arendt (1958, p. 37).

The public realm . . . was reserved for individuality; it was the only place where men could show who they really and inexchangeably were. It was for the sake of this chance, and out of love for a body politic that it made it possible to them all, that each was more or less willing to share in the burden of jurisdiction, defence and administration of public affairs.[60]

Against this we may set the view of Greek politics found in the work of M. I. Finley. Asking why the Athenian people claimed the right of every citizen to speak and make proposals in the Assembly, yet left its exercise to a few, he finds that 'one part of the answer is that the *demos* recognised the instrumental role of political rights and were more concerned in the end with the substantive decisions, were content with their power to select, dismiss and punish their political leaders'.[61] Elsewhere he writes, even more explicitly: 'Then, as now, politics was instrumental for most people, not an interest or an end in itself.'[62] Contrary to what Arendt suggests, the possession or the possibility of exercising a political right may be more important than the actual exercise. Moreover, even the exercise derives its value from the decisions to be taken. Writing about the American town assemblies, Arendt argues that the citizens participated 'neither exclusively because of duty nor, and even less, to serve their own interests but most of all because they enjoyed the discussions, the deliberations, and the making of decisions'.[63] This, while not putting the cart before the horse, at least places them alongside each other. Although discussion and deliberation in other contexts may be independent sources of enjoyment, the satisfaction one derives from *political* discussion is parasitic on decision-making. Political debate is about what to *do*—not about what ought to be the case. It is defined by this practical purpose, not by its subject-matter.

Politics in this respect is on a par with other activities such as art, science, athletics or chess. To engage in them may be deeply satisfactory, if you have an independently defined goal such as 'getting it right' or 'beating the opposition'. A chess player who asserted that he played not to win, but for the sheer elegance of the game, would be in narcissistic bad faith—since there is no such thing as an elegant way of

---

[60] Arendt (1958, p. 41).
[62] Finley (1981, p. 31).
[61] Finley (1976, p. 83).
[63] Arendt (1973, p. 119).

losing, only elegant and inelegant ways of winning. When the artist comes to believe that the process and not the end result is his real purpose, and that defects and irregularities are valuable as reminders of the struggle of creation, he similarly forfeits any claim to our interest. The same holds for E. P. Thompson, who, when asked whether he really believed that a certain rally in Trafalgar Square would have any impact at all, answered: 'That's not really the point, is it? The point is, it shows that democracy's alive . . . A rally like that gives us self-respect. Chartism was terribly good for the Chartists, although they never got the Charter.'[64] Surely, the Chartists, if asked whether they thought they would ever get the Charter, would not have answered: 'That's not really the point, is it?' It was because they believed they might get the Charter that they engaged in the struggle for it with the seriousness of purpose that also brought them self-respect as a side effect.[65]

IV

I have been discussing three views concerning the relation between economics and politics, between the market and the forum. One extreme is 'the economic theory of democracy', most outrageously stated by Schumpeter, but in essence also underlying social choice theory. It is a market theory of politics, in the sense that the act of voting is a private act similar to that of buying and selling. I cannot accept, therefore, Alan Ryan's argument that 'On any possible view of the distinction between private and public life, voting is an element in one's public life.'[66] The very distinction between the secret and the open ballot shows that there is room for a private–public distinction within politics. The economic theory of democracy, therefore, rests on the idea that the forum should be like the market, in its purpose as well as in its mode of functioning. The purpose is defined in economic terms, and the mode of functioning is that of aggregating individual decisions.

At the other extreme there is the view that the forum should be completely divorced from the market, in purpose as well as in institutional arrangement. The forum should be more than the distributive

---

[64] *Sunday Times*, 2 November 1980.
[65] Cf. also Barry (1978, p. 47).   [66] Ryan (1972, p. 105).

totality of individuals queuing up for the election booth. Citizenship is a quality that can only be realized in public, i.e. in a collective joined for a common purpose. This purpose, moreover, is not to facilitate life in the material sense. The political process is an end in itself, a good or even the supreme good for those who participate in it. It may be applauded because of the educative effects on the participants, but the benefits do not cease once the education has been completed. On the contrary, the education of the citizen leads to a preference for public life as an end in itself. Politics on this view is not *about* anything. It is the agonistic display of excellence,[67] or the collective display of solidarity, divorced from decision-making and the exercise of influence on events.

In between these extremes is the view I find most attractive. One can argue that the forum should differ from the market in its mode of functioning, yet be concerned with decisions that ultimately deal with economic matters. Even higher-order political decisions concern lower-level rules that are directly related to economic matters. Hence constitutional arguments about how laws can be made and changed, constantly invoke the impact of legal stability and change on economic affairs. It is the concern with substantive decisions that lends the urgency to political debates. The ever-present constraint of *time* creates a need for focus and concentration that cannot be assimilated to the leisurely style of philosophical argument in which it may be better to travel hopefully than to arrive. Yet within these constraints arguments form the core of the political process. If thus defined as public in nature, and instrumental in purpose, politics assumes what I believe to be its proper place in society.

[67] Veyne (1976) makes a brilliant statement of this non-instrumental attitude among the elite of the Ancient World.

REFERENCES

Ainslie, G. (1982) 'A behavioral economic approach to the defense mechanisms', *Social Science Information* 21, 735–80.
Arendt, H. (1958) *The Human Condition*, Chicago: University of Chicago Press.
Arendt, H. (1973) *On Revolution*, Harmondsworth: Pelican Books.

Arrow, K. (1963) *Social Choice and Individual Values*, New York: Wiley.

Arrow, K. (1973) 'Some ordinal-utilitarian notes on Rawls's theory of justice', *Journal of Philosophy* 70, 245–63.

Asch, S. (1956) 'Studies of independence and conformity: I. A minority of one against a unanimous majority', *Psychology Monographs* 70.

Barry, B. (1978) 'Comment', in S. Benn *et al.* (eds.), *Political Participation*, Canberra: Australian National University Press, pp. 37–48.

Berlin, I. (1969) *Two Concepts of Liberty*, Oxford: Oxford University Press.

d'Aspremont, C. and Gevers, L. (1977) 'Equity and the informational basis of collective choice', *Review of Economic Studies* 44, 199–210.

Downs, A. (1957) *An Economic Theory of Democracy*, New York: Harper.

Elster, J. (1978) *Logic and Society*, Chichester: Wiley.

Elster, J. (1979) *Ulysses and the Sirens*, Cambridge: Cambridge University Press.

Elster, J. (1983a) *Explaining Technical Change*, Cambridge: Cambridge University Press; Oslo: Universitetsforlaget.

Elster, J. (1983b) *Sour Grapes*, Cambridge: Cambridge University Press.

Finley, M. I. (1973) *Democracy: Ancient and Modern*, London: Chatto and Windus.

Finley, M. I. (1976) 'The freedom of the citizen in the Greek world', reprinted as Ch. 5 in M. I. Finley, *Economy and Society in Ancient Greece*, London: Chatto and Windus 1981.

Finley, M. I. (1981) 'Politics', in M. I. Finley (ed.), *The Legacy of Greece*, Oxford: Oxford University Press, pp. 22–36.

Føllesdal, D. and Hilpinen, R. (1971) 'Deontic logic: an introduction', in R. Hilpinen (ed.), *Deontic Logic: Introductory and Systematic Readings*, Dordrecht: Reidel, pp. 1–35.

Goldman, A. (1972) 'Toward a theory of social power', *Philosophical Studies* 23, 221–68.

Haavelmo, T. (1970) 'Some observations on welfare and economic growth', in W. A. Eltis, M. Scott and N. Wolfe (eds.), *Induction, Growth and Trade: Essays in Honour of Sir Roy Harrod*, Oxford: Oxford University Press, pp. 65–75.

Habermas, J. (1982) Diskursethik—notizen zu einem Begründingsprogram. Mimeographed.

Hammond, P. (1976) 'Why ethical measures need interpersonal comparisons', *Theory and Decision* 7, 263–74.

Hansson, B. (1970) 'An analysis of some deontic logics', *Nous* 3, 373–98.

Hirsch, F. (1976) *Social Limits to Growth*, Cambridge, Mass.: Harvard University Press.

Hirschman, A. (1982) *Shifting Involvements*, Princeton: Princeton University Press.

Hogarth, R. M. (1977) 'Methods for aggregating opinions', in H. Jungermann and G. de Zeeuw (eds.), *Decision Making and Change in Human Affairs*, Dordrecht: Reidel, pp. 231–56.

Janis, I. (1972) *Victims of Group-Think*, Boston: Houghton Mifflin.

Kant, I. (1795) *Perpetual Peace*, in H. Reiss (ed.), *Kant's Political Writings*, Cambridge: Cambridge University Press.

Kelly, J. (1978) *Arrow Impossibility Theorems*, New York: Academic Press.

Kolm, S.-C. (1977) *La transition socialiste*, Paris: Editions du Cerf.

Kolm, S.-C. (1981a) 'Altruismes et efficacités', *Social Science Information* 20, 293–354.

Kolm, S.-C. (1981b) 'Efficacité et altruisme', *Revue Economique* 32, 5–31.

Lehrer, K. (1978) 'Consensus and comparison. A theory of social rationality', in C. A. Hooker, J. J. Leach and E. F. McClennen (eds.), *Foundations and Applications of Decision Theory*. Vol. 1: *Theoretical Foundations*, Dordrecht: Reidel, pp. 283–310.

Lipsey, R. G. and Lancaster, K. (1956–7) 'The general theory of the second-best', *Review of Economic Studies* 24, 11–32.

Lively, J. (1975) *Democracy*. Oxford: Blackwell.

Lyons, D. (1965) *Forms and Limits of Utilitarianism*, Oxford, Oxford University Press.

Margalit, A. (1983) 'Ideals and second bests', in S. Fox (ed.), *Philosophy for Education*, Jerusalem: Van Leer Foundation, pp. 77–90.

Midgaard, K. (1980) 'On the significance of language and a richer concept of rationality', in L. Lewin and E. Vedung (eds.), *Politics as Rational Action*, Dordrecht: Reidel, pp. 83–97.

Mill, J. S. (1859) 'Bentham', in J. S. Mill, *Utilitarianism*, London: Fontana Books (1962), pp. 78–125.

Niebuhr, R. (1932) *Moral Man and Immoral Society*, New York: Scribner's.

Parfit, D. (1981) 'Prudence, morality and the prisoner's dilemma', *Proceedings of the British Academy*, Oxford: Oxford University Press.

Pateman, C. (1970) *Participation and Democratic Theory*, Cambridge: Cambridge University Press.

Pattanaik, P. (1978) *Strategy and Group Choice*, Amsterdam: North–Holland.

Rawls, J. (1971) *A Theory of Justice*, Cambridge, Mass.: Harvard University Press.

Riker, W. and Ordeshook, P. C. (1973) *An Introduction to Positive Political Theory*, Englewood Cliffs, N.J.: Prentice Hall.

Runciman, W. G. and Sen, A. (1965) 'Games, justice and the general will', *Mind* 74, 554–62.

Ryan, A. (1972) 'Two concepts of politics and democracy: James and John Stuart Mill', in M. Fleisher (ed.), *Machiavelli and the Nature of Political Thought*, London: Croom Helm, pp. 76–113.

Samuelson, P. (1950) 'The evaluation of real national income', *Oxford Economic Papers* 2, 1–29.

Schelling, T. C. (1980) 'The intimate contest for self-command', *The Public Interest* 60, 94–118.

Schotter, A. (1981) *The Economic Theory of Social Institutions*, Cambridge: Cambridge University Press.

Schumpeter, J. (1939) *Business Cycles*, New York: McGraw-Hill.

Schumpeter, J. (1961) *Capitalism, Socialism and Democracy*, London: Allen and Unwin.

Sen, A. K. (1970) *Collective Choice and Social Welfare*, San Francisco: Holden–Day.

Sen, A. K. (1976) 'Liberty, unanimity and rights', *Economica* 43, 217–45.

Sen, A. K. (1979) 'Utilitarianism and welfarism', *Journal of Philosophy* 76, 463–88.

Sobel, J. H. (1967) ' "Everyone", consequences and generalization arguments', *Inquiry* 10, 373–404.

Smullyan, R. (1980) *This Book Needs No Title*, Englewood Cliffs, N.J.: Prentice Hall.

Tocqueville, A. de (1969) *Democracy in America*, New York: Anchor Books.

Tversky, A. (1981) 'Choice, preference and welfare: some psychological observations', paper presented at a colloquium on 'Foundations of social choice theory', Ustaoset (Norway).

Williams, B. A. O. (1973) 'A critique of utilitarianism', in J. J. C. Smart and B. A. O. Williams, *Utilitarianism: For and Against*, Cambridge: Cambridge University Press, pp. 77–150.

Veyne, P. (1976) *Le pain et le cirque*, Paris: Seuil.

Zeldin, T. (1973) *France 1848–1945*, Vol. 1, Oxford: Oxford University Press.

5. An historical materialist alternative to welfarism*

JOHN E. ROEMER

Modern theories of justice, fairness and social choice are predominantly welfarist: the social objective function that is maximized is
a function of individual utilities only.[1] Many problems arise in the
welfarist tradition, the most celebrated of which is summarized in
Arrow's impossibility theorem (1953).[2] Within welfarism, a position
exists which might be called one of hegemonic individualism: the
postulate that interpersonal utility comparison is impossible or not
meaningful. I call this hegemonic individualism because its philosophical foundation appears to be the belief in the inscrutability of individuals. This position, and its consequences for social choice theory, is
summarized by Arrow:

> To the extent that individuals are really individual, each an auto-
> nomous end in himself, to that extent they must be somewhat

* Added in proof, January 1985: This paper was written in late 1980. Since then, my
opinions on these questions have changed substantially, in part due to work which has
since been done presenting alternatives to welfarism, and clarifying the criticisms of
welfarism towards which I grope here. I cannot now amend this paper adequately to
reflect this, but some abbreviated indications are offered in the first paragraph of
Section VI. For my current views on exploitation and non-welfarist theories of justice,
see Roemer (1985a, 1985b, 1985c).
  I thank Zvi Adar, G. A. Cohen, Leif Johansen, Serge-Christophe Kolm, Thomas
Natsoulas, Amartya Sen, and the participants of the Ustaoset conference for their many
valuable comments, and the John Simon Guggenheim Memorial Foundation for
support.
[1] Welfarism is defined by Amartya Sen, and his criticisms of the concept are
developed in Sen (1977, 1979a, 1979b). The pioneering work of neoclassical social
choice theory, which is welfarist, is Arrow (1953). Harsanyi's (1979) work is welfarist;
the theory of fairness in the economics literature is also welfarist. (Fairness was initially
suggested by Foley [1967], and has been studied by Varian [1974, 1975] and others.)
Virtually all of neoclassical economics, stemming as it does from the utilitarian
approach, is welfarist in so far as it claims to have a theory of social welfare.
[2] Two surveys of the family of results related to the Arrow theorem are Sen (1977) and
Plott (1976).

mysterious and inaccessible to each other. There cannot be any rule that is completely acceptable to all. There must, or so it now seems to me, be the possibility of unadjudicable conflict, which may show itself logically as paradoxes in the process of social decision-making.[3]

Various writers have located the source of the problem of the Arrow impossibility theorem in the hegemonically individualist approach. Harsanyi (1979) insists that utility functions should be cardinal and interpersonally comparable and then shows that the impossibility result dissolves. In Sen's many writings on the subject, he has been sharply critical of welfarism. While he locates the problem of the Arrow theorem in the poverty of information on individual utility functions which is admissible,[4] he continues by 'disputing the acceptability of welfarism *even when* utility information is as complete as it can possibly be'.[5] While Harsanyi resolves the social choice problem by remaining a welfarist, Sen is critical of welfarism even with very complete utility information. Rawls's theory (1971) is also an attempt to escape welfarism by replacing utility functions with some measure of access to primary goods and replacing existing individuals with their surrogates, who make decisions behind a veil of ignorance before they are encumbered with endowments of resources and skills and preferences. Rawls does, however, insist on contractarianism, roughly, the principle that the justice of a situation is decided upon by the people concerned. This will be commented upon later.

This paper outlines an alternative to the welfarist evaluation of the allocation of economic goods which I think follows from the theory of historical materialism. The approach is normative. In its non-welfarism this proposal shares much with Rawls's emphasis on primary goods and Sen's emphasis on basic capabilities (1980). What is peculiar to historical materialism is the concept of exploitation, and hence I begin by presenting a taxonomy of types of exploitation which are central to historical materialism. The historical materialist moral imperative is

---

[3] Arrow (1973, p. 263).

[4] Sen (1979a, p. 544): 'It is this peculiarity of traditional welfare economics in insisting on both that social judgments be based on utility information and that the utility information be used in a particularly poor form, that can be seen as paving the way to inconsistency or incompleteness—and thus to impossibilities.'

[5] *Ibid.*, p. 547.

then discussed, which concerns the elimination of exploitation(s), and the paper continues by following up some questions which are raised by this approach.

For reasons that will be explained, I prefer to claim that historical materialism is concerned with the self-actualization of man and of men, which may not be the same thing as a concern with justice. Although there may be an historical materialist theory of justice, it is not presented here.

I. A taxonomy of exploitation[6]

The approach I will take to evaluate alternative distributions of goods will not be completely general, but historically specific. The discussion will be confined to feudalism, capitalism, socialism, and communism, and ideal types of those models at that. Generality will also be limited in the number of goods which are supposed to be relevant; there are only two, income and leisure. This simplifying assumption is made because I think the main questions of distributional evaluation during the historical period under discussion can be posed by recognizing only these two goods. While some additional justification of this choice of goods will be attempted below, the basis for it is that during the historical period of scarcity under discussion, the main constraint on man's self-actualization is his access to material goods (food, shelter, health care), and income is the proxy for the measure of how binding this constraint is. In particular, the evaluative criterion proposed here will lose its cogency, should an historical era of general abundance ever be reached. (I will not here debate whether abundance can be defined in an absolute way.)

An individual will be said to be better off in state X than in state Y if he receives more income and at least as much leisure in X as in Y. First, this is an extremely weak criterion, as it does not permit comparison of states in which an individual possesses two bundles of income and leisure which are not comparable as vectors. Nevertheless, I will argue that even this weak criterion takes us a long way in evaluating actual history. Second, I do not enquire into the type of work which the individual has to perform in X or Y to get his income. For example: one might argue (and many do) that the self-employed shopkeeper who

6 A fuller discussion of the taxonomy of exploitation is presented in Roemer (1982).

works 80 hours a week for an income of $200 is better off than the factory worker who works 60 hours a week for an income of $210, but I will not. I should like to claim that such examples are of fairly minor significance in the large historical picture.

Rather than speak of the justice of an allocation, the terminology will be of exploitative and non-exploitative allocations. The normative evaluation of alternative distributions will involve assessing the existence of exploitation in those distributions, the ethical imperative being to eliminate exploitation of various types. The relevant type of exploitation, for evaluative purposes, will change with the mode of production; below, exploitations corresponding to feudalism, capitalism and socialism are defined. The presumption behind this taxonomy is that the distribution of income is determined in large part by the property relations which exist, and the main forms of property relations in the period under discussion may be conveniently summarized as feudal, capitalist, and socialist. There are, of course, different distributions possible within one form of property relation, and this will be discussed briefly.

1. *A definition of exploitation.* In every society there is inequality. Not all inequality will be viewed as exploitative, but the notion of exploitation will involve inequality in a way which changes as societies evolve. The inequality of master and slave was viewed (at least by the masters) as non-exploitative in ancient society, as was the inequality of lord and serf in feudal society, although today we consider both of these relationships exploitative. Similarly, Marxists view the inequality in the capitalist–worker relationship as exploitative, although this inequality is conceived of as non-exploitative by many people in capitalist society today. What device can be proposed which distinguishes exploitative from non-exploitative inequality, in a way which is normatively relevant, in a given historical period?

I will say a group is exploited if it has some conditionally feasible alternative under which its members would be better off. Precisely what the alternative is is unspecified for the moment. If two people disagree on whether a particular group is exploited in some situation, then this device leads one to ask: Are they specifying the alternative for the group differently? Different specifications of the alternative will be proposed which generate different definitions of exploitation.

Formally, the alternative will be specified by defining a game 'played' by coalitions of agents in the economy. To define the game, we specify what any particular coalition can achieve on its own, if it withdraws from the economy. The alternative to participating in the economy is for a coalition to withdraw and take what it can achieve on its own, under the specified definition of the game. If a coalition can do better for its members under the alternative of 'withdrawing', then it is exploited at the actual allocation, under that particular specification of the rules for withdrawal.

To make this less abstract, consider the usual notion of the *core* of private ownership exchange economy. The core is the set of allocations which no coalition can improve upon by withdrawing under these rules: that it can take with it the original endowments of its members. Under these particular withdrawal rules, there are certain portions of income and leisure available to any coalition, and one could say a coalition is exploited if its members are receiving a distribution of income-leisure pairs which can be dominated by a distribution of income-leisure pairs achievable by the coalition on its own, given those withdrawal rules. More generally, if we adopt a different rule of withdrawal, i.e. a different way of specifying the achievable rewards of the various coalitions acting on their own, we will have a different game and a different core. The definition is simply this: exploitation occurs, at a given allocation, if that allocation is not in the core of the game defined by the particular withdrawal specification under consideration. That is, a coalition is exploited if it can 'block' an allocation, under the rules of the game. The core of the game is that set of allocations at which no coalition is exploited.

This device captures the idea that we conceive of exploitation as the possibility of a better alternative. My proposal for what constitutes feudal exploitation, capitalist exploitation, and socialist exploitation amounts to naming three different specifications of withdrawal rules. One can then compare different concepts of exploitation by comparing the different rule specifications which define their respective games.

2. *Feudal exploitation.* I shall not be terribly precise concerning the underlying model of feudal economy. Think of agents with various

endowments, who are engaged in production and consumption under feudal relations. A coalition is feudally exploited if its members can be better off by withdrawing under these rules: the coalition can take with it its own endowments. Thus, non-feudally exploitative allocations are, in fact, precisely the core of the usual private ownership exchange game, as conventionally defined and discussed above. This withdrawal specification is the correct one for capturing feudal exploitation, as it gives the result that serfs are exploited and lords are exploiters, which is the notion we wish to capture.[7] Moreover, non-serf proletarians, for instance, will not be an exploited coalition, under these rules, and so the definition captures only feudal exploitation.

To verify this claim, I will assume that feudal serfs owned their own land.[8] Feudal law required them to work the corvée and demesne, not in order to have access to the family plot, but in spite of this access. Thus, were a group of serfs to be allowed to withdraw from feudal society with their endowments, in which we shall include the family plots, they would have been better off, having the same consumption, but providing no labour for the lord. Withdrawal, under these rules, amounts to withdrawal from feudal bondage. (In fact, it has been argued that many serfs would have been better off if they could have withdrawn from bondage, even without their land; surveillance of serfs was necessary to prevent them from running away to the towns, to which they could presumably carry only their non-land endowments.)

There is, however, a possible counter-argument, which could have been advanced by a 'feudal ideologist': serfs would not be better off, he might say, by withdrawing with their own endowments, because they receive various benefits from the lord which they cannot produce on their own, the most obvious being military protection. Also one might believe the lord possessed certain skills or abilities for the organization of manor life, without which the serfs would be worse off. Indeed, the neoclassical analysis of feudalism views serfs as exchang-

[7] An exploiting coalition is one whose complement is exploited. If the characteristic function of the game is super-additive and the original allocation is Pareto-optimal, then an exploiting coalition will always be worse off by withdrawing under the rules of the game. This and other game-theoretic points are investigated in some detail in Roemer (1982).

[8] This is a gross simplification. Relations of property in land under feudalism were vague, by bourgeois standards. What is important for our purposes is that lords did not own the serf's family plots or the commons.

ing their demesne labour for the services provided by the lord.[9] I think this is a gross mischaracterization of feudalism, but it is not necessary to argue that carefully here.[10] The coerciveness of the lord–serf relationship belies the hypothesis of *quid pro quo*.

Perhaps the most important benefit received by the serfs from the lord was military protection. Could serfs, withdrawing with their own assets, have provided such military protection for themselves? If large enough coalitions of serfs withdrew, the answer is yes, assuming the lord possessed no special skills of military organization. Crucial indivisibilities and 'non-convexities' require us, generally, to enquire into the exploitation status only of sufficiently large groups of agents —groups large enough to take advantage (on their own) of any increasing returns to scale or indivisibilities which exist. Thus, I cannot maintain that an individual serf was exploited, by the specified withdrawal criterion; I can maintain, however, that large coalitions of serfs were exploited. For actual historical purposes, the limitation of the jurisdiction of exploitation to large coalitions is no important restriction. (Notice, we can still speak of small *exploiting* coalitions, as an exploiting coalition is the complement of an exploited coalition.) Each of the historical categories of exploitation provided here will classify the vast majority of people as exploited at an appropriate point in time.

3. *Capitalist exploitation.* To test whether a coalition of agents is capitalistically exploited, a different set of withdrawal rules to define the game is specified. When a coalition 'withdraws', it takes with it its *per capita* share of society's non-human property. That is, a coalition can block a particular allocation if that allocation can be improved upon by the coalition, when the initial endowment of alienable non-human assets is an equal-division, egalitarian endowment of goods (but not skills). While the test for feudal exploitation allows agents to exempt themselves from the feudal bond in constructing the alternative against which a current allocation is judged, the test for capitalist exploitation allows agents to exempt themselves from the capitalist bond in constructing the hypothetical alternative, namely, relations of private property in the means of production.

---

[9] See North and Thomas (1973).
[10] But see Brenner (1976, p. 35) for comment on the North–Thomas argument.

Given this phrasing of the alternative, it is not surprising that capitalist exploitation, as here defined, is equivalent to the usual Marxian definition of exploitation in terms of socially necessary labour time and surplus value, a fact which is proved in Roemer (1982). A coalition of agents is Marxian-exploited if the labour value embodied in the goods it can purchase with its revenues from production is less than the labour its members expend in production, the difference being 'surplus value' which is appropriated by other agents. These coalitions can be proved to be precisely the coalitions which are capitalistically exploited, in the sense here defined (under certain special but commonly made assumptions, such as that the technology exhibits constant returns to scale) in standard models of economies.[11]

Just as the feudal ideologist argued that, in fact, serfs would not have been better off had they withdrawn with their own endowments, so a 'bourgeois ideologist' would argue that those who are Marxian-exploited (that is, whose surplus value is appropriated by others) would not, in fact, be better off were they to withdraw with their *per capita* share of society's alienable goods. The surplus value which workers contribute to the capitalist is, in fact, a return to a scarce skill possessed by him, necessary for organizing production (for instance). The bourgeois ideologist's argument is in theory a correct one: if, in fact, equalization of alienable assets would not be sufficient to make Marxian-exploited workers better off on their own, then they are not capitalistically exploited. This is a non-trivial bone of contention between Marxist and bourgeois thinkers, and I will come back to it.

There is, however, also a much less subtle disagreement. A common neoclassical position, I believe, is that exploitation (whatever that means) does not exist at a competitive equilibrium, because everyone has gained from trade as much as possible. How can one say A is exploiting B if A has voluntarily entered into trade with B? Now the model of Marxian exploitation which has been referred to, shows that gains from trade and Marxian exploitation are not mutually exclusive. The proletarian gains from trading his labour power, since otherwise he starves, but his surplus labour time is nevertheless expropriated. What is at issue here is precisely the difference between feudal and

---

[11] For discussion of the inadequacies of the classical Marxian surplus value definition of exploitation, see Roemer (1985a). There are conditions under which the classical definition and the property relations approach outlined here are not equivalent.

capitalist exploitation. The statement that no coalition can gain further from trade amounts to saying that the allocation is in the core of the feudal game: no group of agents, withdrawing with its own endowments, can trade to a superior allocation for its members. Hence this variant of the neoclassical position says 'There is no feudal exploitation under capitalism', a statement rendered true by the well-known fact that competitive equilibria lie in the core of the private ownership game.[12]

To bring more precision into discussions of this nature, it is convenient to differentiate between entrepreneurs and coupon-clippers among the class of capitalists. Entrepreneurs presumably earn a high return on their inalienable endowments, while coupon-clippers earn a return only on their alienable endowments. If we conceive of the capitalist class as predominantly composed of the former, then the statement that 'exploitation does not exist under capitalism' can be consistently interpreted as referring to capitalist exploitation; if the latter, then that statement can only refer to feudal exploitation.

4. *Socialist exploitation.* I now pose the rules of the game which define socialist exploitation. Some endowments were not hypothetically equalized in formulating the game to test for capitalist exploitation: endowments of inalienable assets, skills. Let us picture a socialist economy where private property is not held in alienable assets, but inalienable assets are still held by individuals. Assume the distribution of income reflects this, and those with scarce skills are better off than the unskilled. This inequality is not, of course, capitalist exploitation, as no one holds any private alienable property, and so no coalition can improve its position by withdrawing with its *per capita* share of society's alienable means of production. (Formally, either assume that alienable means of production are unnecessary and non-existent, or that each individual is already assigned his *per capita* share of them in the socialist economy.) We may, however, wish to refer to this inequality as socialist exploitation, characterized as follows. Let a coalition withdraw, taking with it its *per capita* share of *all* endowments, alienable and inalienable. If it can improve the position of its

[12] This well-known theorem of general equilibrium theory can be found, for instance, in Varian (1978).

members, then it is socialistically exploited at the allocation in question.

There are, of course, formidable problems in actually carrying out this procedure, as has been discussed by parties to the Rawlsian debate over talent pooling. (See, for example, Kronman [1979].) How can talents be pooled without destroying them, and so on? I shall discuss this problem below; for the moment, I assume skills can be effectively shared to perform this hypothetical test.

Thus, if we conceive of all individual endowments as being of either the alienable or inalienable type, then an allocation is free of socialist exploitation precisely when it is in the core of the game in which agents withdraw, taking with them equal, *per capita* shares of society's entire bundle of assets and skills. Note how a certain classical conception of socialism and communism is reflected in this definition. The historical task of the socialist revolution is to bring about a regime where each labours according to his ability and is paid according to his work, while the communist revolution (from socialism) transforms the formula so that each is paid according to his need. Thus socialist exploitation is to be expected under socialism: the elimination of differential rewards to ability is not socialism's historical task, only the elimination of differential reward to property ownership. The communist revolution is the one that eliminates socialist exploitation. (More on this below.)

In summary, agents are feudally exploited if they can improve their income-leisure bundles by withdrawing from the economy, exempting themselves only from relations of bondage but not from relations of private property. An agent or coalition of agents is capitalistically exploited if it can improve its income-leisure lot by withdrawing with its *per capita* share of the alienable assets of society and its own inalienable assets (rather than its own assets). A coalition is socialistically exploited if it can improve its lot by withdrawing not only its *per capita* share of alienable assets, but its *per capita* share of inalienable assets.

Briefly, the historical materialist claim is that feudal exploitation of serfs exists under feudalism and is eliminated by the capitalist revolution, capitalist exploitation of proletarians exists under capitalism and is eliminated by the socialist revolution, and socialist exploitation of the unskilled exists under socialism (and perhaps is eliminated by the communist revolution).

5. *Socially necessary exploitation.* An important but tacit assumption was made in performing the hypothetical tests for the various forms of exploitation. It was assumed that when a coalition withdraws, the incentive structure which its members face in the alternative economy set up by the coalition does not differ from the incentive structure in the original economy. In general, this is false. Consider proletarians under early capitalism. Had they been able to withdraw with their *per capita* share of the alienable assets of society, they may very well not have worked long enough to make the income that they had as proletarians; instead they might have chosen to take much more leisure and less income. But even if one strengthens the very weak criterion for judging welfare improvements that I have given, and allows the judgment that these independent producers are better off than the erstwhile proletarians because their leisure-income bundle is superior (even though not strictly dominating), there is a key dynamic consideration. Assuming capitalist property relations were necessary to produce technical innovations in the early period of capitalism, then the coalition which has withdrawn will soon fall behind the capitalist society because of the lack of incentive to innovate. Thus, even the proletarians under capitalism will eventually enjoy an income-leisure bundle superior to the bundle of independent Utopian socialists who have retired into the hills with their share of the capital, assuming enough of the benefits of increased productivity pass down to the proletarians, as has historically been the case.[13] Thus, a more precise phrasing of the criterion is: *were* a coalition able to preserve the same incentive structure and, by withdrawing with its *per capita* share of alienable assets, thereby improve the lot of its members, then it is capitalistically exploited in the current allocation. If, however, the incentive structure cannot be maintained, and in addition, as a consequence, the coalition will be worse off—if not immediately, then 'soon'—then I will say the capitalist exploitation which it endures is socially necessary.

A parallel remark applies to socialist exploitation. We can think of

---

[13] Lindert and Williamson (1980) claim, on the basis of new statistical evidence, that workers in the cities of mid-nineteenth-century England were better off, in terms of real wages, life expectancy, and other quality-of-life indicators, then rural unskilled labourers of 1780. Their work challenges the conventional Marxist view that the 'satanic mills' were an unmitigated disaster for the working class over the period of the industrial revolution. (It is nevertheless true that at any *point* in time until the 1880s or so life expectancy was lower for urban workers than for rural workers.)

the egalitarian distribution of income as the one in which there is no socialist exploitation. (Since the test for socialist exploitation involves hypothetically giving each agent his *per capita* share of both alienable and inalienable assets, it is equivalent to giving him his *per capita* share of total income, not just wealth, assuming all assets are either alienable or inalienable.) However, the incentive structure may be such that, under the egalitarian distribution, the skilled do not have the incentive to use or produce their skills. Hence, in fact, socialist exploitation (which is to say the differential remuneration of skills) is socially necessary. The parallel can be stated more precisely: in early capitalism, differential remuneration to owners of capital was socially necessary, for otherwise capital would not have been produced, and in early socialism, differential remuneration to owners of skills is necessary for otherwise skills will not be produced.

Notice there is a difference between maintaining that capitalist exploitation exists, but is socially necessary, and the argument of the bourgeois ideologist that capitalist exploitation does not exist. For he is maintaining that the capitalists' profits are a return to scarce skills they possess, and hence the income losses workers would suffer in the coalition, when they withdraw with their *per capita* share of alienable assets, are not due to incentive problems, but rather to their lack of accessibility to the skills of capitalists. The bourgeois ideologist claims that the workers under capitalism are experiencing socialist exploitation, not capitalist exploitation. This is quite different from maintaining that capitalists possess no particular skills, but that the regime of private ownership relations in the means of production produces certain behaviour (of competition leading to innovation) which would be absent without those relations. In the second view, it does not matter who the capitalists are, but the workers will be better off if *someone* is a capitalist. Capitalism is socially necessary, but particular capitalists are not!

I will review the idea of socially necessary exploitation. In the initial test performed to determine whether a coalition is exploited (under some set of rules), a *ceteris paribus* assumption was made that the coalition could preserve the incentives of the original society in the coalition. But, in fact, if the coalition adopts different property relations in its coalitional sub-economy, incentives may change in such a way as to alter production possibilities. In relaxing this *ceteris paribus* assumption, I am relaxing the assumption that we can view the game as a cooperative one. Total production in the coalition may, in fact, be a function of income distribution in the coalition. The original *ceteris paribus* test for exploitation is one of technological feasibility only; whether exploitation, when it exists, is socially necessary is a question of social feasibility. How will behaviour alter with changes in property relations?

II. The historical materialist imperative

According to historical materialism, feudal, capitalist and socialist exploitation all exist under feudalism. At some point feudal relations become a fetter on the development of the productive forces, and they are eliminated by the bourgeois revolution. Feudal exploitation is outlawed under capitalism. Although the proletarians might not immediately appear to be better off than the serfs, because of the non-comparability of their income-leisure bundles, one can assert that quite rapidly (in a generation, perhaps) they are better off owing to the rapid development of the productive forces and the increase in the real wage. Thus, feudal relations were eliminated when they became dynamically socially unnecessary, in the sense that the incentive structure of feudalism was less efficient than the alternative incentive structure (without feudal relations, but in which private property is respected) in producing technical innovations which soon made the exploited better off. Under capitalism, only capitalist and socialist exploitation continue to exist. Capitalist exploitation in the beginning is socially necessary, as I have discussed; eventually, however, it becomes a fetter on the development of productive forces.[14] Large coalitions of proletarians could be better off by withdrawing with their *per capita* share of the non-human means of production, and organizing themselves in a socialist way, because capitalism no longer performs a progressive innovating function, compared to what society is capable of accomplishing under socialist organization. Thus, capitalist exploitation becomes socially unnecessary, and is eliminated by socialist revolution. Under socialism, capitalist exploitation is outlawed, but

---

[14] See Brenner (1976) in which the backwardness of French agriculture is blamed on a premature victory of the independent peasantry against agricultural capitalism, as contrasted with the English experience.

socialist exploitation exists and, most would argue, is socially necessary, at least in the early period.[15]

Historical materialism, in summary, claims that *history progresses by the successive elimination of dynamically socially unnecessary forms of exploitation.* In addition, I maintain its ethical imperative is this: distribution X is better than distribution Y if there is less socially unnecessary exploitation in X than in Y. (I do not here explore a precise meaning of 'less': in practical cases, I think the decision is usually clear-cut.) The best one can hope for is to eliminate exploitation which is socially unnecessary at a given time.

An example might be worthwhile to demonstrate how this criterion can be applied even to compare income distribution in two societies with the same property relations, say capitalist ones. Let us suppose that at this point in history, capitalist property relations are socially necessary in both the United States and Norway. (If they were eliminated, incentives would so dissipate that the currently capitalistically exploited workers would be worse off.) Suppose, however, a large group of American workers (say 100,000,000 of them) could withdraw, taking with them their share of the capital stock and resources, and instead of setting up socialism, set up a capitalist regime like Norway's, allowing private ownership of the means of production, but with more progressive redistribution than exists in the United States. Perhaps this system of property relations would be sufficient to coax out the incentives necessary to maintain productivity and innovation (as it is in Norway), and the coalition will all be better off than under United States capitalism. We would then say there is socially unnecessary capitalist exploitation in the United States, although not all of the capitalist exploitation is socially unnecessary. Perhaps, however, there is no socially unnecessary capitalist exploitation in Norway—any more radical redistribution would destroy incentives. Either, then, the ethical imperative calls for moving to socialism, if the incentive structure can bear that transformation, or it does not, but further redistribution within capitalist relations is not at issue.

Why should it be considered good to eliminate socially unnecessary exploitation? There appear to be two quite different reasons, neither of which is an appeal to egalitarianism as a fundamental goal.

---

[15] This issue, and a concrete discussion of existing socialist countries, is pursued more extensively in Roemer (1982).

Although the successive elimination of exploitations, as described, does move society in the egalitarian direction (though not necessarily monotonically), as it requires the successive removal of categories of property relations (first feudal property, then capitalist property, then socialist property), I do not think it is necessary to found the argument for the ethical imperative of the theory on an appeal to egalitarianism as an unquestionable good.

The unquestionable good is the self-actualization of men and of man. I cannot provide a definition of what it means to become self-actualized; suffice it to say, the idea exists not only in Marx, but in modern social theorists.[16] Rawls views primary goods as essentially distinguishable from other goods in that they are necessary for the realization of life plans, an ideal similar to self-actualization of *men*. Sen argues for including the notion of 'basic capability equality' in theories of distributive justice:

> It is arguable that what is missing in all this framework is some notion of 'basic capabilities': a person being able to do certain basic things. The ability to move about is the relevant one here, but one can consider others, e.g., the ability to meet one's nutritional requirements, the wherewithal to be clothed and sheltered, the power to participate in the social life of the community. The notion of urgency related to this is not fully captured by either utility or primary goods.[17]

Now in the historical period of scarcity (from feudalism to at least the present day), the major block to the self-actualization of men is the constraint they face in providing themselves with material goods needed for survival. Hence, if a coalition is exploited in a socially unnecessary way, its ability to self-actualize (or realize its basic capabilities) can be improved upon. But, the social aggregator might reply, by eliminating the exploitation of the underdog, perhaps the basic capabilities of the former exploiters will be infringed upon. I would argue historically this has not been the case. Typically, in the historical period under discussion, distribution is characterized by a large class of exploited, whose lack of access to income sharply hinders

---

[16] For empirical work on self-actualization see Maslow (1973).
[17] Sen (1980, p. 218).

their self-actualization, and a small class of exploiters, who have income above the level that is necessary to provide them with the opportunities for self-actualization which are available at the current level of development of culture and the productive forces. (I do not think one needs to be a millionaire to reach the threshold of material constraint on self-actualization today; one needs only an income a little more than mine.) Thus, distributions are better if they allow more people access to the material goods necessary for self-actualization, or basic capabilities, and the elimination of socially unnecessary exploitation is the best one can do in approaching that goal.

This is the first argument. Notice that there is no essential distinction between the self-actualization of men and Rawls's opportunity for constructing life plans or Sen's basic capability achievement. The second argument, however, distinguishes self-actualization from these other two concepts.

Marx and Engels did not commend capitalism simply because it gave the direct producers higher income, but because it broadened their outlook, through the introduction of science and culture and, in a word, the development of the productive forces.[18] The development of the productive forces is the necessary condition for dynamic self-actualization of *man*. Or the development of the productive forces is the proxy for the self-actualization of man. This type of dynamic self-actualization is to be distinguished from the fulfilling of basic capabilities or life plans, because presumably any given person would rather have the resources necessary to realize his life plan or realize his basic capabilities, but no individual can be dynamically self-actualized: for the dynamic self-actualization of man is not something which happens to the individual, but rather to the species as it develops culturally over time. There is a judgment being made, that man who understands how the universe works is more self-actualized than man who does not. The agnosticism of much contemporary social science is so pervasive that it would not agree with such a judgment.

To summarize, the elimination of dynamically socially unnecessary exploitation is the relevant evaluative criterion, because, first, during the era of scarcity, it increases the opportunities for short-run self-actualization of *men* by providing the formerly exploited large group with access to the wherewithal for basic living, and second, according

[18] See G. A. Cohen (1974) for discussion of this point.

to historical materialism, the elimination of socially unnecessary exploitation is necessary for the development of the productive forces, which is the proxy for self-actualization of man.[19] That the elimination of dynamically socially unnecessary exploitation is necessary for the development of the productive forces is a claimed theorem of historical materialism. (Its classical statement is that each form of property relations, or economic structure in G. A. Cohen's terminology, comes to fetter the development of productive forces, and will therefore be cast aside.)[20]

In bourgeois (or, more precisely, middle-class) society, an opinion has developed in the last decade asserting that the era of abundance has already arrived—in the terms relevant here, that the constraints on self-actualization are no longer material, but are cultural or psychological. I refer to the proponents of no-growth, small-is-beautiful, and the like. This might appear to nullify the appropriateness of the criterion for self-actualization. I do agree, and did state, that the criterion only applies in the historical period of scarcity, when men's self-actualization is hindered by their access to material goods, given the degree of self-actualization possible during that historical period due to the development of culture at the time. Only the most myopic view, however, can maintain that such constraints are no longer binding on the people of the world in the late twentieth century. It is surely true that such a material constraint is not binding on many people in the industrialized capitalist countries. That, however, is not an argument for slowing down growth, but for continuing growth with redistribution to those billions who are severely materially constrained. Moreover, there have been recurring examples in history of revulsion against the material excesses of contemporary society, by small minorities of people, frequently taking the form of religious cults which withdraw to an ascetic alternative. The modern no-growth movement is not a new response to fears of population growth and resource depletion.

These arguments do not locate the historical materialist imperative in consideration of justice. It is not claimed there are no such ethical imperatives; simply, I am not prepared to present the argument. The

---

[19] I owe the distinction between the self-actualization of man and of men to G. A. Cohen.

[20] For a detailed study of the claims of historical materialism see Cohen (1978).

problem can be seen in the debate concerning justice in Marxism. Wood (1972, 1979) presents broad evidence that Marx considered justice to be a superstructural concept, therefore corollary to extant property relations. In this reading, an income distribution is just if it conforms to the laws of society. Income from slavery is just in a slave society and unjust in a capitalist society. Wood (1979) says: 'Marx prefers to criticize capitalism directly in terms of this rational content [crisis, anarchy of production, etc.] and sees no point in presenting his criticisms in the mystified form they would assume in a moral ideology.' There is a strong undercurrent of adherence to Wood's position in Marxism, which can be seen in the frequent insistence that Marxism criticizes capitalism from an objective and not normative point of view. Contrarily, Husami (1978), Elster (1980), and Cohen (1981) maintain that Marxism claims that capitalism is unjust, *inter alia*. (Another approach is Brenkert [1979], who claims: 'Marx's critique cannot be one of justice but can be and is one of freedom.') Cohen (1981) presents the intuitively compelling position that the passion with which Marxism indicts capitalism can only come from a belief in its injustice, and cannot be sustained simply by the belief in the technical fact that its period of historical usefulness is past.

Wood's claim is that the Marxian concept of justice provides no effective criterion for criticism of society. If, however, Marxism does entail a less relativist and more powerful concept of justice, which is the intuitively plausible position, I think it will permit statements such as this: although early capitalism was progressive, and its exploitation was socially necessary, it was unjust. A concept of justice will permit the existence of a necessary evil. I have claimed that historical materialism indicates a direction in history, and an accompanying imperative based on self-actualization, which is not necessarily the same imperative as one based on justice.

Should such a concept of justice exist consonant with historical materialism, it will differ, in particular, from Rawlsian justice in this way. For Rawls, any inequality that maximizes the potential of those most constrained in their possibilities to achieve life plans is just. If capitalist relations of property are progressive in the Marxian sense in early capitalism, then early capitalism is just for Rawls. Rawlsian justice is determined by evaluating the allocation in question against currently feasible counterfactuals—feasible in the sense of incentive considerations. A Rawlsian just allocation is, in the present language, one in which the worst-off coalition suffers no socially unnecessary exploitation. The identification of such an allocation with justice flows from the contractarian approach, in which groups in society hypothetically agree to certain forms of organization because they can do no better. If a Marxian theory of justice exists which can perform the task mentioned, it cannot be contractarian in this sense.

III. Desert, return, and surplus

The terms 'desert', 'return', and 'surplus' often occupy central roles in theories of distribution and economic justice. In particular, many interpretations of the Marxian theory see Marxism as mandating some distribution either according to deserts or needs. (See Sen [1973, Ch. 4]; Elster [1980]; Husami [1978].) Needs will be discussed in the next section.

When one speaks of the return to a factor, there are two possible meanings: that the remuneration to that factor is a just one (or a deserved one) or that it is a necessary one. If we wish desert and return to connote two different things, then return must be taken to mean *necessary* return. That is, remuneration to a factor is a return if, without that remuneration, production cannot take place, or that factor cannot be reproduced. This is the predominant usage of the term in economic theory. Return is associated with economic law: if the payment to factors of their returns is not made, economic law is violated, and the economy cannot function. If the factor capital does not receive its return of profits, then the economy will not reproduce, capital will not reproduce. Wages are the necessary return to labour, profits to capital: such is economic law.

The historical materialist theory outlined implies something quite different. Whether or not profits should be viewed as a return to capital depends on whether capitalist property relations are necessary for the reproduction of capital. In early capitalism, as I have argued, profits are properly viewed as a return to capital, in that such remuneration to capital is socially necessary. Had the capital stock been socialized, it would not have been reproduced at that stage in the development of the productive forces and the consequent possible social organization and consciousness of the workers. In late capitalism, however, profits are no longer properly viewed as a return to capital. Rather the statement is the following: if profits are not returned to capital, then capital will not reproduce itself, *given the constraint* that the capitalist property relations be maintained. It is, however, fallacious to conclude that, therefore, capitalist property relations are necessary. For example, when conservatives argue for a decrease in corporation profit taxes, to encourage more creation of capital, liberals have but one reply: that lowering the tax rate will not encourage more investment. Marxists, on the other hand, might admit that decreasing taxes will increase investment, but take this not as a mandate to decrease taxes, but rather as an indictment of the system of property relations in which such forms of investment stimulus are needed. Investment 'strikes' do not prove the necessity of lowering the tax rate but of socializing capital, if private property in the means of production is no longer socially necessary (in the technical sense we have defined).

Even if profits are a return to capital, it is fallacious to say they are a return to particular capitalists. As has been argued, no individual capitalist need be reproduced to reproduce capital, even when capitalism is socially necessary. This situation differs from the relationship between the worker and his wage, because of the inalienability of labour power from the person. The wage is a return to the worker, as well as to labour power.

There are two possible meanings of surplus. The more conventional meaning is this: a surplus is a portion of income which exists beyond what is consumed by the direct producers, at the moment in question. I think, on the contrary, it is more useful and precise to contrast surplus with return, as the above definition does not, as follows: surplus is that part of the economic product which does not have to go to any particular factor for that factor to be reproduced. When capitalism is no longer socially necessary in the dynamic sense, then profits become a surplus. In early capitalism, profits were not a surplus, because it would not have been possible for society to do with them what it pleased, by hypothesis. A surplus is something extra, not in the sense of extra over present consumption, but in the sense that it need not go to any particular factor for economic reproduction. Under this definition, the neoclassical contention that profits are a return and not a surplus is sensible, though (Marxists claim) wrong. The neoclassical argument, as with the tax example, is frequently myopic in taking property relations as fixed. Profits are surely not a surplus, in the second sense, if one refuses to consider the possibility of socialism.

Similarly with differential remuneration to skill under socialism. When such payments are socially necessary (for those skills to be reproduced), then they are properly viewed as returns to skill. When they are no longer socially necessary, then such remuneration is not a return to skill, but a surplus. Economic law evolves as society evolves.

The term desert connotes justice, and since no theory of justice is here proposed, there can be no definition of desert. We can say some things do not appear to be deserts, however, on the basis of the intuitions evolved for necessary features of an historical materialist theory of justice. If early, progressive capitalism is unjust (as proposed), then profits, although a return, are not a desert. (Perhaps a clue to justice is this: justice must attach itself to judgments of the role of persons, not of things. Even when profits are a return to capital, because they are not a return to capitalists, they cannot be deserved.) It is somewhat more difficult to decide whether a skill remuneration is a desert, at least when such remunerations are socially necessary, if we stipulate, as I did not in previous discussions, that skills are embodied inalienably in individuals, at least at a given moment. Under those conditions, such a payment is a return to skill and to the skilled person, since for that skill to continue to exist (at least in the short run), he must receive the payment. When skill wage differentials become socially unnecessary, they no longer constitute a return to skill, and I would argue, therefore, surely not a desert, since socially unnecessary inequality should not be just (whatever justice means).

IV. Needs

To discuss the problem of needs using the framework of the taxonomy of exploitation which has been proposed, imagine the vector of endowments of a person living under feudalism as consisting of four different types of component:

$$\Omega^i = (\Phi^i, \pi^i, \sigma^i, \nu^i)$$

Person i has, first, a component ϕ^i which measures his degree of feudal privilege. (We might measure this in such a way that the sum of ϕ^i over all persons i is zero. Positive feudal privilege is the property owned by the lord and negative feudal privilege is the serf's feudal property.) π^i is the vector of private property in the means of production, σ^i is the vector of labour skills possessed by i, and N^i is the vector of needs of i. A component in this endowment vector will be positive when the person is well-off (as in the feudal component), so the needs component is specified in this fashion: a large negative component of v^i means person i has a lot of some need. Alternatively stated, the vector v^i is the freedom from various needs enjoyed by the individual. As in the feudal component, the needs components might be normalized by assigning a person with a normal need a zero endowment of that need type.

This identification of separate components for needs and skills is too facile, in fact, for needs are surely not independent of skills. The form that excessive neediness frequently takes is the inability to work. Needs whose major effect on individual welfare is through the deprivation of skill will already be captured in the vector of skill endowments, and are not independently identified in the needs vector. The needs vector should only list features which are independent of skill: some people may require more income to treat a disease, although the disease does not affect their skill; or large family size, though not affecting skill, may require more income. This dichotomy is not always so neat. Nevertheless, a first approximation to the problem of needs can be made in this way.

The taxonomy of exploitation can be stated in this notation. To test whether an agent or coalition S of agents is feudally exploited, it withdraws taking an endowment of vector of the form:

$$(0, \sum_S \pi^i, \sum_S \sigma^i, \sum_S v^i)$$

That is, the coalition is allocated its *per capita* share of feudal privilege (which is to say none, since by hypothesis the total endowment of feudal privilege is zero), and its own private property of everything else. If it can improve the lot of its members under this allocation, then the coalition was feudally exploited, with the usual *ceteris paribus* assumption concerning incentives.

To assess whether a coalition is capitalistically exploited, it takes with it, in the hypothetical withdrawal, its *per capita* share of both feudal privilege and alienable property, that is:

$$(0, \frac{S}{N} \underset{N}{\Sigma} \pi^i, \underset{S}{\Sigma} \sigma^i, \underset{S}{\Sigma} \nu^i)$$

but keeps its own skills and its own needs. (N stands for the universe of agents and S for the coalition in question. N and S also stand for the cardinalities of these coalitions.) The test is then performed to see whether the coalition can produce an allocation superior for its members to what they were getting in the original economy.

To assess whether a coalition is socialistically exploited, it is hypothetically given endowments in which it receives its *per capita* share of feudal privilege, alienable property, and skills, but retains its own needs. This will give rise to an egalitarian distribution of income: that is, the egalitarian distribution of income is the one in which socialist exploitation has been eliminated.

We may still consider that the needy are exploited in an egalitarian distribution, and might name this needs exploitation. A coalition is needs-exploited if it could improve its lot (from the egalitarian distribution) by withdrawing with its share of all assets and its share of needs. That is, with the endowment vector:

$$(0, \frac{S}{N} \underset{N}{\Sigma} \pi^i, \frac{S}{N} \underset{N}{\Sigma} \sigma^i, 0)$$

If the coalition can arrange a superior allocation for its member under this allocation of endowments, then it was needs-exploited in the egalitarian distribution. To be less abstract, consider a person with high needs, whose ν^i components are therefore negative. When he withdraws with his share of needs (that is, with zero needs), he will be better off than before—because he will have the same income as before (namely, his *per capita* share of total income, since the distribution of all other assets is egalitarian), but fewer needs than before. Similarly, a person who is less needy than average will be worse off when he withdraws with his share of every component of the endowment, as he will have the same income as in the egalitarian distribution but more needs.

The elimination of needs exploitation requires compensating people for their needs. A distribution in which needs exploitation is absent is one in which the needy receive more income than the needless. Thus, the passage to communism (defined as 'from each according to his ability, to each according to his needs') requires the elimination of two kinds of exploitation which exist in early socialism: first, socialist exploitation, and second, needs exploitation. It is generally assumed in the Marxist literature that socialist exploitation will be eliminated first, but there appears little reason to suppose this will be so. Modern societies seem more willing to compensate the specially needy than the specially unskilled. Observe also that there is no name for that hypothetical mode of distribution in which socialist exploitation is eliminated but not needs exploitation. (In socialism both forms of exploitation exist and in communism, neither does.)

The theory of historical materialism appears most plausible in explaining the elimination of feudal and capitalist exploitation, that is, in reference to the bourgeois and socialist revolution. Whether it is useful in the study of other pre-capitalist modes of production or post-socialist transitions is problematical. That is: can it be claimed that property rights in skills, as they exist in socialism in their limited sense (of entitling the skilled to higher income but not to capitalize that income), will eventually come to fetter the development of productive forces, thus 'mandating' the abolition of socialist exploitation? What will be the crises of socialism in which such fettering becomes manifest? Or, alternatively, if socialist exploitation is to be eliminated, will that occur as a conscious action of society, to further self-actualization or justice, even though the development of the pro-ductive forces is not the issue? The same questions can be put for needs exploitation. These are issues which cannot be seriously discussed with our historical experience. Suffice it to say, historical materialism is a theory which interprets at most two historical transitions, and I can conceive of no argument so convincing as to enable us to claim today that the same mechanism will accomplish the elimination of socialist and needs exploitation.

An asymmetry appears to exist in the way society eliminates feudal and capitalist exploitation, contrasted with how one might imagine it will eliminate socialist and needs exploitation. When feudal exploi-tation is abolished, that form of property is eliminated. When capitalist exploitation is abolished, capitalist property is owned by no one. In contrast, if skills and needs are truly inalienable, then those forms of property cannot be socialized or eliminated: rather, the elimination of the relevant form of exploitation can only occur through compensation of the unskilled or the needy. Entitlements to the fruits of skill (and unneediness) can be changed by society, but the physical possession of those assets and liabilities cannot be (or so it now seems). Some writers take this as evidence that socialist exploitation can never be eliminated, because of problems considered to be intractable in the pooling of talents (see Kronman [1979]), a conclusion which does not follow. Although talents, perhaps, cannot be pooled, it is not necessary to do so to eliminate socialist exploitation, but only to pool income from the exercise of those talents—a difficult problem, but not metaphysically insurmountable.[21]

It may be the case that the elimination of socialist exploitation will be accomplished by the elimination of the socialist forms of property in this way: that all people become equally skilled and hence the socialist property *differential* disappears. Certainly skill differentials under capitalism are highly accentuated compared with what they will be in a society in which special opportunities due to wealth and status have been eliminated. (Or, perhaps it is more important to emphasize the inverse: in which the special alienation and waste of human potential of those lacking wealth and status is eliminated.) A similar remark applies to the socialization of needs.

Historical materialism, then, claims that history progresses by successively eliminating the entitlements of individuals to the fruits from various forms of property, in a certain order. The *reason* a property entitlement becomes eliminated is that it comes to fetter the development of the productive forces. The *mechanism* through which the property entitlement in question is abolished is class struggle, where classes are defined with reference to entitlements respecting the income from the property in question. (Why this mechanism develops and succeeds is beyond our scope.) The *good* accomplished in this process is the self-actualization of men and of man; perhaps justice is also approached, although only perfunctory remarks concerning that issue have been made. Whether the *reason* will continue to apply

---

[21] For discussions of talent equalizing distributional mechanisms published since this text was written, see Dworkin (1981) and Roemer (1985b).

beyond the capitalist era is unclear; whether the argument for the *good* (of self-actualization, though perhaps not of justice) will continue to apply when all societies eliminate the material constraints on immediate survival is also unclear.

V. Some comparisons with Rawlsian justice

One difference between the Rawlsian conception of justice and the ideas presented here has been noticed above. A situation in which the only exploitation that exists is socially necessary would, I believe, be called just by Rawls. I do not think the historical materialist approach should render that decision.

A second difference arises from Rawls's attempt to found his difference principle (the maximin distribution) in a contractarian mechanism. The analogous approach for this paper would be to provide a contractarian basis for using the self-actualization of man and of men, which has not been done. In that sense, Rawls's theory appears to go one step closer to founding the theory in the decisions of the individuals involved than does this historical materialist approach.

In another paper, R. E. Howe and I have attempted to evaluate the consistency of Rawlsian contractarianism.[22] In the Rawlsian theory, the distributional problem can be thought of as arising from the incentive structure, which, it is assumed, characterizes behaviour. If primary goods are distributed in a very egalitarian way, total production may be much less than if the distribution of primary goods is skewed. (Think of wealth as the only primary good, and endow persons with a capitalist mentality; a highly skewed distribution of wealth may produce much larger total production than an egalitarian one.) Total production may be so much higher that everyone can be made better off with some inequality in the original distribution of primary goods. In particular, if there is no incentive problem in motivating people to work, then the maximin distribution of wealth (the primary good) will be the egalitarian distribution. Rawls's paradigm is thus only interesting because there *is* an incentive problem, and thus some inequality will be accepted in the social contract in the original position. Howe and I show, in a model of the Rawlsian theory, that the following three tenets or requirements cannot logically co-exist:

[22] R. E. Howe and J. E. Roemer (1981).

(1) a model of individuals who react according to the incentive structure which provides the tension in Rawls's model;

(2) an objective function in which persons or souls evaluate their package of *primary goods* (as opposed to their *utility*);

(3) a *contractarian* theory, where judgment is passed upon the justice of an allocation by individuals and coalitions of individuals who make up society, according to a criterion of the variety described in (2).

There may be other models of the Rawlsian theory in which the inconsistency demonstrated in our model does not appear. If, however, the Rawlsian inconsistency is robust, I would tentatively resolve the problem by not insisting on contractarianism as part of the theory. (Incentive problems cannot be assumed away, and I endorse the non-welfarist approach of an emphasis on primary goods.) For this reason, no attempt has been made in this paper to ground self-actualization as an outcome of a contractarian mechanism.

Another inconsistency of the Rawlsian theory has been remarked upon by Plott (1978), but I do not support his criticism. Plott proves an impossibility result for the Rawlsian theory of justice. He takes seriously the aggregation problem of preferences over primary goods. Suppose there is more than one primary good. Then how can we decide between two bundles of primary goods, neither of which dominates the other, component-wise? Plott solves the problem by endowing individuals with preference orderings over primary goods. This done, he is able to push the problem into the Arrow mould, and prove an impossibility result: that if there are at least three primary goods, then there must exist an individual who is a dictator, in the sense that any change in the distribution of primary goods which he prefers will necessarily improve the lot of the 'least advantaged'. That is, there is an individual who dictates what the maximin allocation of primary goods is.

I think a Rawlsian objection to Plott's model would be that he has let in utility functions through the front door. Although actors do not use their utility functions to evaluate all goods, they use them to evaluate primary goods. Perhaps Rawls thinks that the distribution of primary goods is too important to be left to the subjectivity of idiosyncratic tastes. He also appears to dispense with the index problem of weighting primary goods by assuming they are usually tied to each other (if you have more wealth, you have more power and more freedom), so that for all practical purposes there is no problem in comparing two actual packages of them.[23] Thus I believe Plott's model goes against the non-welfarist spirit of Rawls. In this paper, I have avoided the Plott problem by imposing an ordering on leisure-income bundles.

Three key differences between the Rawlsian approach and this paper's are: that inequalities which are socially necessary may still be called exploitative, whereas they may be Rawlsian-just; no philosophical foundation for the self-actualization criterion has been claimed, in contrast to the Rawlsian attempted basing of the difference principle in a contractarian mechanism; the historical specificity of the approach outlined here.

VI. Final remarks

Classical social choice theory is welfarist, in admitting information only in respect to the utilities of agents. There are, however, 'resourcist' theories of social welfare (Rawls, Sen (1980), Dworkin (1981)) which admit information not only on the utilities agents receive, but on the resources they get. Resources may matter for reasons other than the utility they produce, as the agents themselves report or measure it. The setting of classical social choice theory is too informationally impoverished to permit a discussion of resource allocation: there are no 'economic environments' which give rise to social states, taken as primitives, over which agents have preferences. Resourcist theories of justice begin with an informationally denser domain, consisting of economic environments, in which resources, primary goods, needs, and preferences of agents are specified. I have argued that historical materialism and its corollary sequence of exploitations are linked with the resourcist, non-welfarist goal of self-actualization. A formal approach to resourcist social welfare theory is proposed in Roemer (1985c).

During the historical era of scarcity, I claim the self-actualization of men is furthered by the elimination of socially unnecessary exploitation. First, that elimination always relaxes the binding material constraints on survival for some people, without lowering others (the

[23] Rawls (1971, pp. 93–4).

erstwhile exploiters) to the level at which they become so materially constrained. Secondly, according to historical materialism, the elimination of such forms of exploitation is necessary for the development of the productive forces, which radically reduces the material constraints on the self-actualization of men. Thus, elimination of exploitation fosters dynamic efficiency, which is more important than the static equity it brings. And thirdly, the development of the productive forces leads to self-actualization of man, since it can be taken as the proxy for man's understanding of the world and his ability to act upon it.

Little has been said concerning how people think, in various societies, about the forms of inequality which exist. The theory of exploitation, as proposed, claims to illuminate historical development without reference to the perceptions of people making the history. Historical materialism should imply a corollary of sociology of moral beliefs or injustice, in which exploited classes come to think of themselves as exploited and hence rebel against their exploitation. Perhaps the exploited learn to classify feudal bondage as exploitative under feudalism, but not capitalist inequality; capitalist inequality becomes viewed as exploitative under capitalism, but not inequality due to skill differentials, and so on. Each mode of production might inculcate the beliefs in the exploited class which are necessary for it to perform its 'historical task'. Or, perhaps the exploited are less discriminating in the kinds of inequality they struggle against (viz. the egalitarian experiments which have appeared throughout history), but the only kind of inequality which can be successfully eliminated at a given point in time is the one which is not then socially necessary.

Indeed, why call the inequality due to differential ownership of alienable productive assets capitalist exploitation, when it is socially necessary? Why not call it capitalist inequality, and reserve the term exploitation for those forms of inequality which have become socially unnecessary? (Then, we would say proletarians are not capitalistically exploited under early capitalism.) Does my terminology not simply replace 'inequality' with 'exploitation', thus excising what should be an important linguistic distinction? I have not previously raised this question, because the proper response requires, I think, a sociology of injustice which has not been developed.[24] The answer tentatively

[24] A sociology of injustice or moral belief is not necessarily the same as a theory of justice.

proposed is: people in a society come to view certain forms of inequality as exploitative when the possibility, and eventually the necessity, arises to eliminate them. This occurs before those forms of inequality become socially unnecessary. The choice to describe early capitalist inequality as exploitative is made to respect the nascent moral beliefs of the proletarians, who come to view such inequality as exploitative. The bourgeois revolutionaries of 1789 would speak of feudal exploitation but capitalist inequality; the proletarian revolutionaries of 1848 would speak of capitalist exploitation, even before it became socially unnecessary; the proletarian revolutionaries of 1917 would speak of capitalist exploitation and socialist inequality; and there are those in current socialist society (witness the recent history of China) who would refer to the phenomonon as socialist exploitation, even before the differential remuneration of skills has become socially unnecessary in the dynamic sense. Paying such respect to the participators in class struggles of the time is not done here out of sentimentality, but to suggest the link, which must be made, between the theory of exploitation and the theory of class struggle.

That is to say: the theory of exploitation exposited here is a development of only one facet of historical materialism, what is called, by both friend and foe, its technological determinism. The other facet, here neglected, is the mechanism historical materialism proposes to realize its determinist prediction, class struggle of the exploited against the exploiters. It is the sociology of injustice, which will be an historical materialist theory of preference formation, which must provide the link between these two facets, and I wish to suggest the necessity for that link by the chosen terminology and these remarks.

REFERENCES

Arrow, K. (1953) *Social Choice and Individual Values*, New Haven: Cowles Foundation Monograph.

Arrow, K. (1973) 'Some ordinalist-utilitarian notes on Rawls' theory of justice', *Journal of Philosophy* 70, 245–63.

Brenkert, G. G. (1979) 'Freedom and private property in Marx', *Philosophy and Public Affairs* 8, 122–47.

Brenner, R. (1976) 'Agrarian class structure and economic development in pre-industrial Europe', *Past and Present* 70, 30–73.

Cohen, G. A. (1974) 'Marx's dialectic of labor', *Philosophy and Public Affairs* 3, 235–61.

Cohen, G. A. (1978) *Karl Marx's Theory of History: A Defence*, Princeton: Princeton University Press.

Cohen, G. A. (1981) 'Freedom, justice and capitalism', *New Left Review*, 3–16.

Dworkin, R. (1981) 'What is equality? Part II: equality of resources', *Philosophy and Public Affairs* 10, no. 4, 283–345.

Elster, J. (1980) 'Exploitation and the theory of justice', History Institute: University of Oslo.

Foley, D. K. (1967) 'Resource allocation and the public sector', *Yale Economic Essays*, Vol. 7.

Harsanyi, J. (1979) 'Bayesian decision theory, rule utilitarianism, and Arrow's impossibility theorem', *Theory and Decision* 11, 289–318.

Howe, R. E. and Roemer, J. E. (1981) 'Rawlsian justice as the core of a game', *American Economic Review* 71, 880–95.

Husami, Z. (1978) 'Marx on distributive justice', *Philosophy and Public Affairs* 8, 27–64.

Kronman, A. (1979) 'Talent pooling', New Haven: Yale Law School (forthcoming *Nomos*).

Lindert, P. H. and Williamson, J. G. (1980) 'English workers' living standards during the Industrial Revolution: A new look', Department of Economics Working Paper No. 156, University of California: Davis.

Maslow, A. H. (1973) 'Self-actualizing people: A study of psychological health', in R. J. Lowry (ed.), *Dominance, Self-Esteem, Self-Actualization: Germinal Papers of A. H. Maslow*, Monterey, Calif.: Brooks/Cole Publishing Co., 177–202.

North, D. and Thomas, R. (1973) *The Rise of the Western World*, Cambridge: Cambridge University Press.

Plott, C. (1976) 'Axiomatic social choice theory: An overview and interpretation', *American Journal of Political Science* 20, 511–96.

Plott, C. (1978) 'Rawls' theory of justice: An impossibility result', in H. W. Gottinger and W. Leinfellner (eds.), *Decision Theory and Social Ethics, Issues in Social Choice*, Dordrecht, Holland: D. Reidel Publishing Co.

Rawls, J. (1971) *A Theory of Justice*, Cambridge: Belknap Press.

Roemer, J. (1982) *A General Theory of Exploitation and Class*, Harvard: Harvard University Press.

Roemer, J. (1985a) 'Should Marxists be interested in exploitation?' *Philosophy and Public Affairs* 14, no. 1, 30–65.

Roemer, J. (1985b) 'Equality of talent', *Economics and Philosophy* 1 (forthcoming).

Roemer, J. (1985c) 'The mismarriage between bargaining theory and distributive justice', Department of Economics, University of California, Davis Working Paper no. 253, forthcoming in *Ethics*.

Sen, A. K. (1973) *On Economic Inequality*, Oxford: Oxford University Press.

Sen, A. K. (1977) 'On weights and measures: Informational constraints in social welfare analysis', *Econometrica* 45, 1539–72.

Sen, A. K. (1979a) 'Personal utilities and public judgments: Or what's wrong with welfare economics?' *Economic Journals* 89, 537–58.

Sen, A. K. (1979b) 'Utilitarianism and welfarism', *Journal of Philosophy* 76, 463–89.

Sen, A. K. (1980) 'Equality of what?' in Sterling McMurrin (ed.), *The Tanner Lecture On Human Values*, Cambridge University Press and University of Utah Press.

Varian, H. (1974) 'Equity, efficiency, and envy', *Journal of Economic Theory* 9, 63–91.

Varian, H. (1975) 'Distributive justice, welfare economics, and the theory of fairness', *Philosophy and Public Affairs* 4, 223–47.

Varian, H. (1978) *Microeconomic Theory*, New York: Norton.

Wood, A. (1972) 'The Marxian critique of justice', *Philosophy and Public Affairs* 1, 244–82.

Wood, A. (1979) 'Marx on right and justice: A reply to Husami', *Philosophy and Public Affairs* 8, 267–95.

6. Interpersonal comparisons: preference, good, and the intrinsic reward of a life

ALLAN GIBBARD

I. From happiness to the satisfaction of preferences

Perhaps the most striking contrast between nineteenth-century treatments of utilitarianism and recent treatments concerns the interpersonal comparability of utility. Whereas nineteenth-century utilitarians recognized no special problem with such comparisons, in the eyes of more recent writers the dubious status of such comparisons has loomed as the central threat to the intelligibility of utilitarianism as a whole. It is not that nineteenth-century utilitarians all regarded utilitarianism as unproblematically intelligible. With his usual care Sidgwick discussed the question of whether pleasures are commensurable, and hence whether 'a sum of pleasures are intrinsically unmeaning' (1907, II.iii).[1] He did that, however, as part of his treatment of egoism; when the time came in his treatment of utilitarianism to discuss the intelligibility of the 'Greatest Happiness', he simply remarked that the subject had been covered before (IV.i.2). He recognized, then, that comparisons of utility are problematic, but recognized no special problem for comparisons that are interpersonal. Recent writers, on the other hand, have tended to regard personal utility as unproblematical and interpersonal comparisons as dubious. Why the change?

The problem of interpersonal comparisons of utility has been primarily an economists' problem. That is not to say, of course, that it has been of concern to economists alone, or that it should be. It is economists, though, who set the problem in the form it now takes for workers in a variety of fields. The way a problem is perceived is often a matter of its history as well as of the nature of its subject-matter, and

[1] References to Sidgwick (1907) are by Book, Chapter, and Section, on the pattern that 'IV.i.2' means Book IV, Ch. i, Sect. 2.

the history of thought about utility in this century has been made chiefly by economists.

Among English-speaking economists, I gather, the historical lore is this. In the old days of Edgeworth and Pigou, utility played two roles in economic theory. In the theory of supply and demand, utility determined choice: an economic agent was postulated always to choose, from among the bundles of goods available to him, a bundle of greatest utility. In welfare economics, utility measured a person's welfare and the proper goal of economic policy was taken to be the maximization of total welfare: the sum of the utilities of those affected by the policy. Then in the 1930s a number of things happened. In the first place, under the influence of logical positivism, English-speaking economists followed the lead of Pareto, and came to deny the meaningfulness of a sum of individual utilities: to deny that my welfare could be added to yours to give a meaningful total. That is to say, they denied the meaningfulness of interpersonal comparisons of differences in utility. The two denials amount to the same thing, for my gain exceeds your loss if and only if their sum is positive (with losses taken as negative gains).[2] In the second place, economists noted that in the theory of supply and demand, quantitative utilities were superfluous. An economic agent, on the old theory, always chooses, from among the alternatives open to him, the one of greatest utility for him. Equivalently, though, it could be said that the agent chooses, from among the available alternatives, the one he most prefers. When the theory of rational choice is put that way, quantitative utility has been purged from it and replaced by preference orderings—which cannot be added together.[3]

That left no gap in the theory of supply and demand, where sums of utilities had never figured, but in welfare economics it left an abyss. Deprived of sums of utilities, the welfare economist had no standard for ranking economic states. A state x could be counted as better than a state y only of x were *Pareto-superior* to y: if someone preferred x to y

---

[2] This is not to say that the two denials are, in the strict sense, logically equivalent. Their equivalence does follow from the assumptions that the same people exist in all the situations being compared, and that all combinations of individual utility levels are possible.

[3] I mean this as 'lore' rather than as careful history, in that I recount my impression of what is thought to have happened—of the picture of history that influences current thought. Robbins's denial (1932, Ch. 6) seems to have had an especially strong impact on English-speaking economists.

and no one preferred y to x. Pareto superiority, however, can rarely be used to settle issues of economic policy, for ordinarily no alternative is Pareto-superior to all others.

To span the resulting theoretical abyss, economists constructed a 'New Welfare Economics' that turned out, indeed, to have considerable content.[4] Many of the results of the Old Welfare Economics, it was discovered, could be put in terms sanctioned by the New. What had been necessary conditions for a greatest feasible total utility became necessary conditions, or even necessary and sufficient conditions, for a Pareto optimum—a feasible state to which no feasible state is Pareto-superior. The New Welfare Economics, though, included no basis for assessments of distribution—for thinking one Pareto optimum more desirable than another. In a choice between two Pareto optima, individual interests conflict, and the New Welfare Economics recognized no factual basis for saying which way the preponderance of interests fell.

The economists of this story made two striking departures from the ways of thinking of such nineteenth-century utilitarians as John Stuart Mill and Sidgwick—departures that may help to explain why interpersonal comparisons of utility became specially problematic in their eyes. In the first place, they applied tough, positivistic standards of intelligibility to the concepts they used. Questions of intelligibility, then, were central in their thinking in a way that such questions had not been for nineteenth-century writers. In the second place, partly as a result of positivistic scruples, they replaced pleasure with preference-satisfaction as the standard of intrinsic value in utilitarian theory. Sidgwick and most other nineteenth-century utilitarians took the total happiness of sentient beings[5] as what was to be maximized, with the happiness of a person defined as the net balance of pleasure over displeasure in his life. The mainstream of economic theorists in the past half-century have taken something else as what was to be maximized: the sum of the degrees to which the preferences of people are satisfied. That has its advantages. Preference has a positivistic respectability that pleasure lacks: we can tell what a person prefers by

---

[4] A discussion of the 'New Welfare Economics' can be found in Little (1957, Ch. 6–9). One textbook presentation is in Varian (1978, Ch. 5).

[5] Though hedonistic utilitarianism must presumably count pleasures and displeasures of beasts as well as people, I shall for the most part speak as if only the welfare of people were at stake.

offering him a choice and seeing what he does.[6] Then too, preference plays a central part in positive economic theory. The concept of an interpersonal sum of degrees of preference satisfaction, though, has had no such recommendation, and so leaders of economic thought have rejected it as nonsense.

What I shall do here is to offer a highly qualified brief for the nineteenth-century view of the problem. The switch to preference-satisfaction as a standard of personal welfare has been a mistake. We should indeed have serious doubts as to the intelligibility of utilitarianism, but the doubts we should primarily have are equally doubts as to the intelligibility of rational egoism or rational prudence. If I can fathom what is best for me, there may be no barrier in principle to my seeing what is best, on balance, for us. There may be no precise sense, however, in asking what is best for me—and whether there is is a pressing question for ethical theory.

II. Anti-paternalism and the satisfaction of actual preferences

A person's well-being is the degree to which his preferences are satisfied: that is the claim we shall be examining at the outset. The standard here is the satisfaction of preferences, not the satisfaction of people. Strictly put, the claim is this: given two possible histories of the universe, x and y, a person j is better off in x than in y if and only if j prefers x to y—if and only if, that is to say, j prefers x's being the actual history of the universe to y's being the actual history of the universe. This claim is far from innocuous, and indeed far less innocuous than it may seem when it is put in terms of 'satisfaction'. Applied to a person, 'satisfaction' suggests the absence of cravings, or a feeling of well-being. Apart from connotations of intensity, there is little contrast between satisfaction in this sense and happiness. It is not satisfaction in this sense, however, that is being set forth as the measure of how well off a person is. The claim is rather that a person is better off the more the world is as he prefers. Now to be sure, were a person's preferences only for feelings of well-being, there would still be little contrast to be

---

[6] Or at least it has been widely accepted among economists that all preferences can, in principle, be revealed by choices in possible experimental situations. Tversky, however, reports experiments that cast serious doubt on this claim.

drawn between satisfying a person and satisfying his preferences. It seems clear enough, though, that people want things other then feelings of well-being: sometimes they want revenge, sometimes posthumous fame, sometimes the fidelity of friends or spouse, sometimes the well-being of others. A jealous husband may even prefer a 'fool's hell' in which his suspicions rage but his wife is in fact faithful, to a 'fool's paradise' in which his suspicions are allayed but in fact he is unknowingly cuckolded. In that case, even though he experiences more satisfaction in the fool's paradise than in the fool's hell, his preferences themselves are more fully satisfied in the fool's hell than in the fool's paradise. According to the claim we are examining, he is therefore better off in the fool's hell than in the fool's paradise.

Why, then, should we take a person's welfare to be, not the degree to which he himself is satisfied or happy, but the degree to which his preferences are satisfied? In welfare economics, that is to some extent a matter of theoretical convenience: it is when preferences do measure welfare that welfare economics yields striking results. That is to say, the striking results come when, given alternative possible courses of events, a person is always taken to be best off in the one he most prefers. To be sure, the classical results of welfare economics do not lose their interest if matters of preference and matters of welfare are distinguished. Those results can still be seen as applying to the special case of economic agents each of whom is a pure rational egoist—who, given any two possible courses of events, always prefers the one in which he is better off. Indeed the classical results apply even more straightforwardly to a kind of constrained rational egoist: to an agent who is a rational egoist within a moral constraint of respect for the laws of property and the tax laws. Once enforcement of the constraints of the market and tax system are taken for granted, though, the results apply most widely if everyone is made a rational egoist by stipulation: if to be better off simply is for the history one prefers to be realized.

The move to take preference as the standard of welfare, though, had a more serious rationale. First, preference, as I have said, is a respectable concept: at least in principle, a preference can be elicited by offering a choice. Mere respectability, though, would not make preference the correct measure of welfare. Height too is a respectable concept, but a person's welfare is not measured by his height. There needed to be more reason to think that what a person prefers is always better for him. That reason was put forward under the banner of 'anti-paternalism': that each person should be free to make his own decisions in matters that primarily affect him alone, that the best judge of what is for the good of a person is always that person himself.

Considerations of anti-paternalism, though, fail to support preference as a measure of welfare. To see this, we need first to distinguish two versions of the view that a person's preferences settle what is best for him. The first is the view that whatever policy, course of action, or the like a person actually prefers is shown by the fact of his preference to be better for him. The second is that of any two alternatives, the one that is better for him is the one he would prefer if he were fully and vividly aware of everything involved. The first is a test by *actual preferences*; the second, I shall say, by *ideally informed preferences*.[7] Now, if anti-paternalism supports either test, it would seem to support the test by actual preferences. Anti-paternalism is the view that a person—or a normal adult, at least—should be free to make his own choices in matters that directly affect only himself. He should be 'free to make his own mistakes'. That means that in self-regarding matters, he should be free to choose his course of action on the basis of his own beliefs, even if those beliefs are mistaken. Others may legitimately try to convince him by argument that his beliefs are mistaken, but if they fail, they may not legitimately coerce him into doing what is best for him. They may not even coerce him to act as he himself would choose to if he had his facts straight.[8] Now if actual preferences settle what is best for a person, that explains why paternalism is illegitimate: the very fact that he chooses a course of action then shows it to be best for him. If what is best for a person is settled by his ideal preferences, on the other hand, that speaks prima facie in paternalism's favour. For paternalism may consist in forcing a person to take the course of action he ideally prefers—what he would prefer, that is, with full and vivid awareness of everything involved—rather than the course of action he actually prefers. There may be good arguments against paternalism

---

[7] This distinction is commonly made; see, for instance, Brandt (1978, p. 249). It can be questioned whether the distinction applies to intrinsic desires, or solely to instrumental desires. In the latter case, what I say here should be read as an attack on using actual instrumental desires as a sure test of a person's good.

[8] Here I follow Mill's classical presentation of anti-paternalism (1859, Ch. 4).

even on the ideal preference theory, but they will be of the indirect sort that are available on other theories of welfare too. If we want a theory that condemns paternalism in an especially straightforward way, it will have to be the actual preference theory.

Properly understood, however, the issue of paternalism is not whether a person's uninformed preferences settle what is best for him. To see this, consider a typical case in which an issue of paternalism arises. Suppose a man has cancer, and thinks that laetril will cure it whereas standard treatments will not. Suppose that in fact, standard treatment would cure it whereas laetril would not. If he is an adult of ordinary understanding then it would, according to such anti-paternalists as Mill, be illegitimately paternalistic to act to prevent him from seeking treatment with laetril. Does that mean that he is better off being treated with laetril than he is being treated the standard way? The anti-paternalist could plausibly be committed to no such thing, and there are much more plausible grounds for rejecting the policy of forced treatment. Note first that the choice is not simply one of whether he shall be treated in the standard way or with laetril; it is whether he shall receive standard treatment against his will or be free to seek the treatment he wants. Now to be sure, every attempt should be made to expand these options, at least if the attempts do not themselves involve the use of force: if the person can be convinced of the advantages of standard treatment, then the option of his freely receiving standard treatment will have been opened. If, though, as much persuasion has been tried as the person will tolerate, and he still remains so dubious of the claims of established medicine and so credulous of the claims of the purveyors of laetril that he prefers the laetril treatment, then the alternatives are down to the two I have given, and a real dilemma of paternalism is unavoidable.

Now when that happens, the anti-paternalist may argue against forced treatment in at least two ways that have some plausibility. In the first place, he may claim that it is better for a person to die miserably under a treatment he chooses than to recover through a forced cure, which he has rejected despite all attempts to persuade him. Perhaps self-government is more worthy of a human being, even if it leads to misery, than the status of a happy child. That is not to say, though, that other things equal, dying miserably under treatment by laetril is better for him than recovering under standard treatment. The person is genuinely mistaken about what is better for him: to say that people should be free to make their own mistakes is not, surely, to say that nothing a person chooses could ever be a mistake.

More plausibly, the anti-paternalist might defend the policy of leaving normal adults free to make their own mistakes as Mill defends it. Mill concedes that in some cases the results of forcing a person to submit to what experts judge to be best for him are better than the results of leaving the person free to choose on the basis of unwarranted beliefs that go against expert opinion. He denies, though, that there can be a satisfactory procedure for deciding when to force upon a person what the experts judge best. He denies, that is, that any such procedure would have more desirable results on the whole than the procedure of letting adults of normal understanding decide for themselves.[9]

The more plausible theory of preference as a test of welfare, then, is the ideal preference theory. That, indeed, accords more with common sense. In the case of social policies, for example, we do not suppose that whatever policy a person prefers is better for him. Would a teenager be better off with a lower minimum wage for teenagers than for adults? That is not, we suppose, simply a matter of which policy he prefers. It may be that with a lower minimum wage he would get a job and with the standard minimum wage he would not. He may not realize this, and so prefer that the standard minimum wage apply to him. At the same time, he may prefer a job at the lower minimum wage to no job at all with the standard minimum wage. It would seem strange to conclude that the standard minimum wage is better for him simply because he prefers it, if his preference is based on a mistake about the consequences of the alternative policies. If he prefers the lower minimum wage with its actual consequences to the standard minimum wage with its actual consequences, it seems more reasonable to conclude that the lower minimum wage is better for him. Ideal preference may be a plausible criterion of welfare; actual preference is not. Whether, and under what circumstances, a person should ever be forced to submit to what is best for him is a separate issue—and that alone is the issue of paternalism.

[9] Mill (1859, Ch. 4, par. 9, 12).

III. Ideally informed preferences

A person's ideally informed preferences are the preferences he would have if he were fully and vividly aware of everything involved. Do a person's ideally informed preferences, then, provide a test of what is more and what is less to a person's good? It would seem not. For my ideally informed preferences may be a result not only of what I judge best for myself, but of what I judge best for other people. With full and vivid awareness of everything involved, I may prefer course of events x to course of events y because I judge x considerably better than y for others, even though I also judge y to be somewhat better for myself than x. If so, my ideally informed preference does not agree with my judgment of what is best for me. Now I see no reason to suppose that in such cases I must be wrong in my judgment of what is best for me—as would have to be the case if my ideally informed preference were always for what is better for me. Such a case, after all, need not be one in which others' being better off makes me better off, say because I take joy in the thought of their happiness and am miserable at the thought of their unhappiness. Even if I am made miserable by the misery of others, that may not be my sole reason for wanting them to be happy. There are two strategies I might pursue to alleviate my own misery: to take steps to relieve theirs, or to take steps to make myself less vividly aware of theirs or sensitive to theirs. The latter will often be the more effective strategy, and it is often the strategy we adopt—but not always. A person may really want others to be less miserable, and not simply want to be made less miserable by their misery. In that case, he may prefer a policy that he knows will leave him less well off than some alternative, simply because it will be much better for others.

Perhaps, though, what matters most directly is not how 'well off' a person is in the ordinary sense. Perhaps what is crucial to ethics is rather the degree to which his ideally informed preferences are satisfied. The case for this might be put as follows. Suppose two alternatives are indifferent in their effects on everyone but me, by every standard that we find at all plausible. Then is it not my preferences that should rule, at least if they are not rooted in ignorance? Who is to overrule my preferences on the matter? On matters that concern me alone, should I not be sovereign? And is that not to say that my preferences, if ideally informed, should rule, even if that conflicts with what is best for me? In general, this line of thought concludes, an ideally informed preference should rule, whenever no one else's ideally informed preference would oppose it.

This claim is difficult to evaluate. Perhaps the best test of its plausibility would be a case in which ideally informed preferences are starkly opposed to considerations of welfare: a case of two possible histories x and y such that everyone is better off in x than in y, but y stands higher than x in the ideally informed preferences of everyone (or as I shall say, everyone *ideally prefers* y to x). Can there be such cases? Not if each person's ideal preferences are welfare-based, in the sense that they are founded exclusively on a positive concern for his own welfare and the welfare of others. Indeed, if individual preferences are welfare-based, then it cannot be that going by welfare, course of events x is Pareto-superior to course of events y, whereas going by ideally informed preferences, y is Pareto-superior to x.[10] For suppose that going by welfare, x is Pareto-superior to y: considerations of his own welfare favour x over y, and no considerations of the welfare of others speak for y over x. No one will ideally prefer y to x, for no considerations of anyone's welfare favour y over x. Thus going by ideally informed preferences, as well as by welfare, x will be Pareto-superior to y.

More bizarre grounds for ideally informed preferences can produce hypothetical cases of the kind we are seeking: cases of conflict between unanimous ideally informed preference and unanimous considerations of welfare. Consider a universe of egalitarians who live on three separate planets with no communications. They are universal egalitarians, in that they each ideally prefer a state E of the universe in which everyone is equally well off to a state B in which there are inequalities of welfare, even if everyone is better off in the unequal state B than in the equal state E. Suppose that in the unequal state B, each planet is homogeneous: within each planet, there are no inequalities of welfare. Thus on each planet, everyone can take satisfaction in that, so far as he knows, everyone is equally well off, and everyone with whom he interacts is equally well off (though he realizes that for all he knows, there may be planets on which people are better off or worse off than the people on his). In state of the universe E, everyone

---

[10] This was proved by Pareto.

in the universe is equally well off, but not so well off as the worst off in state B. In state B, as in state E, each person knows that everybody on his planet is equally well off, and is ignorant of the state of anyone in the rest of the universe.

Here if we think unanimous preferences settle what is best, then we must think state E better than state B: we must think it better for everyone to live in universal equality, though unaware of the fact, than for everyone to be better off and still equally as well off as anyone he knows about. Is that an ethical view that we should hold? Suppose we see no value in an equality that hurts everyone and produces no gains of fraternity or community. Then it is hard to see why we should prefer universe E for these people. For in that case, their preference is based on a moral ideal that we do not share, and whose realization among people who do hold it would accomplish nothing that we can recognize as of value. It seems that in the case of a stark conflict between unanimous preference and unanimous considerations of welfare, unanimous considerations of welfare settle what is better.[11]

IV. Self-interested preference

The discussion so far has hinged on the implausibility of regarding the satisfaction of a person's morally-based preferences as a contribution to his welfare—at least when he takes no satisfaction in thinking his morally-based preferences satisfied. Could we perhaps, though, identify a self-interested component of a person's ideally informed preferences, and take that component to reflect his good? That would, of course, sacrifice the simple behavioural test for informed preference: the test of informing him, offering him a choice, and seeing what he chooses.[12] Suppose, though, we can nevertheless somehow make sense of the concept.[13] Call the self-interested component of his ideally informed preferences his *SI-preferences*, and a utilitarianism that takes a person's SI-preferences as definitive of his welfare *SIP-utilitarianism*. Might it make sense to read a person's 'utility' as the degree to

[11] For a debate over whether preferences all told, even if they reflect both individuals' self-interest and their moral views, can be taken as indicators of welfare; see the symposium by Arrow and Brandt in Hook (1967).

[12] But see footnote 6 above.

[13] Overvold (1980) proposes a criterion for self-interested preference. I shall not attempt here to judge whether his proposal works.

which his SI-preferences are satisfied, and to take his utility, so glossed, as a measure of his good?

The problem with this proposal is that a person's preferences can change, and when a choice he makes will affect his later preferences, then it is unclear which of the preference scales he might have measures the relative worth of the alternatives.[14] Consider a simple, stock example: a young man choosing between a seminary and a secular university. With enough sophistication and insight, he might realize that if he chooses the seminary, he will come to value the life of austere religious contemplation, whereas if he chooses the secular university, he will come to value a rich mixture of sensual and intellectual pleasures. Which preferences count?

We might be tempted to count his SI-preferences at the time of the choice; for is it not they that determine which alternative is in his self-interest at the time of the choice? Perhaps so—but suppose he mistakenly chooses against his SI-preference. Suppose he initially prefers the life of austere religious contemplation, but mistakenly thinks that his commitment to that life will best be fostered by experience in a worldly university. If as a result of his university experience he comes to lead a life he values as he leads it, but would have despised when he initially chose the university, can we conclude that life has been bad for him? If not, his initial SI-preferences do not measure his good or his welfare, and so cannot reasonably be taken as his 'utility' for ethical purposes.

Should we then evaluate his 'utility' at each stage in his life by the SI-preferences he has at that stage? We would then judge his life at the university by the worldly preferences he has developed there, and that seems to make sense. How, though, should we judge the alternative life he might have led at the seminary? Should we judge it by the worldly preferences he actually develops, but would not have developed there? Then at the time of the choice, how much each alternative would be to his good is indeterminate: if he goes to the seminary, then going there is more to his good than going to the university, whereas if he goes to the university, then going to the university is more to his good than going to the seminary.

---

[14] The difficulties raised by changes of intrinsic preference are central to Brandt's attacks on preference-satisfaction utilitarianism (1978, pp. 249–51; 1982). Elster (1979 and 1983) treats various aspects of preference change.

Perhaps, then, we should judge his going to the university by the SI-preferences he has by virtue of going to the university, but judge his going to the seminary by the SI-preferences he would have if he had gone to the seminary. But to try to do so would be incoherent—at least if preference is relational, which it seems to be. A person can prefer one alternative to another. I am quite willing to grant, moreover, that his preference will have an intensity, that it may be great or slight. Can a person, though, 'prefer' a single alternative—except in the sense of preferring it to another that is understood from the context? Can we speak of the 'degree to which he prefers it', or the 'intensity with which he prefers it', without at least implying a particular alternative to which he prefers it? If not, then it makes no sense to compare the degree to which he would prefer the seminary had he gone there to the degree to which he prefers the university having gone there. 'Prefers to what?' the reply must be.

Level of preference, I am saying, cannot be characterized in terms of strength of preference alone. That is to say, a person's level of preference for a single alternative is not fixed by the strength of his pairwise preferences between alternatives. If level of preference is to make sense at all, so that one person's level of preference for an alternative can be compared to another person's, or a person's level of preference for an alternative at one time can be compared with his level of preference for it at another time, we need to make sense of something more than a person's strength of preference at a time for one alternative over another.

This point may be obscured by the use of utility scales to represent strength of preference. When we use utility scales in that way, we may be tempted to say that the young man of our stock example is better off in the university than in the seminary if and only if his utility is greater in the university than in the seminary—but that is no standard at all. The point to realize is that a utility scale that represents direction and strength of preference alone has an arbitrary zero point. Consider the matter: Let $P_j^{zt}(x, y)$ be person j's strength of preference, in history z at time t, for history x over history y. Suppose we want to characterize j's utility, in history z at time t, for history x. Call that $U_j^{zt}(x)$. If utility scale U_j^{zt} is to represent only strength of preference, then the only requirement on the scale is that for any two histories x and y,

$$U_j^{zt}(x) - U_j^{zt}(y) = P_j^{zt}(x, y). \tag{1}$$

Once we have such a scale for each z and t, however, we can add an arbitrary constant $k(z, t)$ for each z and t, and let

$$V_j^{zt}(x) = U_j^{zt}(x) + k(z, t)$$

for each x, z, and t. Requirement (1) will still hold for the new scale V_j^{zt}. We will not, however, have the kind of invariance we need to compare levels of utility in different possible courses of events— namely that

$$V_j^{xt}(x) - V_j^{yt}(x) = U_j^{xt}(x) - U_j^{yt}(y).$$

Unless $k(x, t) = k(y, t)$, it may even be that one side is positive and the other negative. In that case, even which of x and y we take to make j better off will rest on an arbitrary choice of scale, on this proposal.

Strength of preference, then, underconstrains utility, in the sense that a version of utilitarianism based on preference strength alone would be incoherent. It remains possible, though, for all I have said, that further, non-arbitrary constraints on utility scales can be found that would render a preference-strength utilitarianism coherent. Consider the prospects for SIP-utilitarianism. Suppose it turned out that a person's genuinely intrinsic SI-preferences never change, and indeed *would* never change no matter what he or anyone else chose to do. They are independent of time and circumstance. If that were so, then SIP-utilitarianism would yield well-determined prescriptions for all cases in which who comes to exist is not in question. For we could then calibrate the utility scales of a person at different times and in different courses of events by the simple rule that a person's utility for a given possible history is presumed not to depend on time and circumstance. Thus, given invariant preference strength, we could presume invariant preference level. Furthermore, we could then make sense of the 'utility' of a person's coming to exist and leading a certain kind of life. We could take that to be the strength of his SI-preference for existing and leading that life, over never coming to exist. Or at least we could do so on the further assumption that a person is SI-indifferent among all possible histories in which he never comes to exist.

V. The intrinsic reward of a life

In the previous three sections, I have been trying to make ethically pertinent sense of a person's 'utility', and to do so in terms of his preferences of some kind or other. I have been able to do so only by retreating to talk of a 'self-interested component' of his preferences, and by making strong, dubious assumptions about the nature of that component: most notably, that SI-preference is unchanging in the course of a person's life. Now, in the course of this retreat, I have had to abandon the behaviouristic appeal of the kind of preference that can be revealed in simple tests. I have renounced the 'anti-paternalistic' rationale for taking a person's preferences to define his good. If this series of withdrawals has indeed been forced on us, as I think it has, then the upshot, it seems to me, is that we are left with no good reason for wanting to define a person's 'utility' in terms of his preferences—at least for purposes of fundamental ethical theory. It is time to explore alternatives.

As we do so, we must not expect to find something that will be revealed by a simple behavioural test, as choice has been supposed to reveal preference. What can be revealed most simply is choice, and that, I have been saying, has no straightforward bearing on ethics. It would be surprising if the psychological magnitudes that bear most fundamentally on questions of ethics were also those that most directly manifest themselves in crudely observable aspects of behaviour. We may hope that the psychological magnitudes we need for ethics figure in the best empirical psychology, but it will have to be an empirical psychology of considerable refinement.

What, then, is 'utility'? The question should not be what the term means in ordinary language—or, indeed, what any other term means in ordinary language. I have been using the terms 'utility', 'welfare', 'benefit', 'interest', what is 'best for' a person, and the like roughly interchangeably, and that suits my purposes here. The important question of interpersonal comparison of utilities should be conceived as follows. The problem of interpersonal comparisons is, as I have said, pressing only for ethical theory. Moreover, though it may be a problem for any ethical theory that will bear scrutiny, it is most directly a problem for utilitarianism. If the question 'Can there be meaningful interpersonal comparisons of utility?' must be met by the question,

'What do you mean by utility?', the answer to this retort should be the one that makes the original question most significant. Roughly, we should construe 'utility' as whatever it is that utilitarians need to compare among different persons.

The role 'utility' or 'welfare' plays in utilitarianism is this. According to the theory, there is a psychic magnitude that characterizes each person, which is such that the intrinsic value of any course of events is the sum of those magnitudes for the people involved. Our question should be, 'What are those psychic magnitudes?' We need to ask, 'What sorts of interpersonal psychic comparisons must be meaningful for the utilitarian theory of intrinsic value to be meaningful?' Once that is settled, we can go on to ask whether those comparisons indeed are meaningful.

Now at the outset, the question demands a shallow answer. Different theorists who fall under the broad label 'utilitarian' have held different theories of intrinsic value. On Bentham's account, intrinsic value is the total balance of pleasure over displeasure. On Mill's official theory, intrinsic value is also a balance of pleasure over displeasure, but with each pleasure weighted by its 'quality'. In places, though, Mill talks as if intrinsic value is total net 'happiness' in a sense in which knowledge, virtue, and dignity can all be parts of a person's happiness.[15] Sidgwick, like Bentham, settles on a straight balance of total pleasure over total displeasure, but his account of pleasure and displeasure is sophisticated in a way that makes it unclear whether he and Bentham are talking about precisely the same thing. Indeed Sidgwick offers his account as the one in terms of which Mill would have put his objections to Bentham if he had seen matters clearly.[16] Moore, whom we might and might not want to count as a utilitarian, had a theory of value that is clearly inconsistent with Sidgwick's. What are intrinsically good, according to Moore, are organic wholes which may all include pleasure, but which also include 'ideal goods' such as knowledge—things that are not, as pleasure, displeasure, and probably happiness must be, intrinsic qualities of experience.[17] Thus to the extent that utilitarians and quasi-utilitarians have been clear in their theories of intrinsic value, they have held intrinsic value to be the sum of various different psychic magnitudes. It makes no sense, then, to

---

[15] See Bentham (1789) and Mill (1863, Ch. 2).
[16] Sidgwick (1907, II.ii.2, esp. pp. 127–8).   [17] Moore (1912, Ch. 7).

ask what kinds of interpersonal psychic comparisons must be meaningful if the utilitarian theory of intrinsic value is to be meaningful. The answer to that depends on the specific version of utilitarianism we consider.

I propose, nevertheless, to try to make sense of the question in its general formulation. Mill, Sidgwick, and Moore, I want to say, were engaged in a common enterprise—one which we too can attempt. This claim is partly historical, at least for Mill and Sidgwick: they had, in some respects, a common guiding rationale for their ethical theorizing. The claim is also reconstructive: moves they made are intelligible and significant when seen against the guiding rationale I shall impute to them. That rationale, I should say at the outset, is one I think worthy of careful and respectful consideration, and perhaps even of adherence, but it is also highly problematical.

Utilitarians have sought a systematic and fundamental basis for criticizing and evaluating moral convictions and ethical thought in general. They have held that pre-philosophically, our moral convictions and our standards for criticizing and evaluating moral convictions do not constitute a coherent, satisfying whole; if they seem to, careful dialectic will reveal that they do not. Even where matters seem clear to one person, there will often be no general agreement on them.[18] A person who wants to reflect fully about ethics, then, needs to ask himself whether the things about ethics that seem clear to him are validated in a way that ensures that those who disagree are making a mistake—and that if anyone else disagreed, he too would be making a mistake.

What is needed, then, is a fundamental basis for criticizing and evaluating moral convictions and ethical thought in general. The basis should be neutral among people. It should be one it would be reasonable for anyone to accept, and indeed that would be accepted by any sufficiently reasonable person who thought enough about the problems involved. Perhaps, of course, all this is too much to demand, and if that turns out to be the case, we shall have to think where that leaves the problem of finding a rational basis for ethical reflection. First, though, the search should be made.

[18] The culmination of this utilitarian tradition is Sidgwick's criticism of the view that the morality of common sense provides axioms for a system of normative ethics (1907, III.ii–xi).

One way to proceed with the search is to ask why morality matters. It seems to matter intrinsically, but if we are uncertain about what morality requires, we are then uncertain what it intrinsically matters to do. Is there a rationale for resolving this uncertainty? If so, the apparently unexplainable importance of the demands of morality is explainable after all. Now a possible strategy of search is a sceptical one. Suppose morality does not matter after all, and see what goes wrong. Suppose everyone pursues his own goals, uninhibited by moral constraints. What would be wrong with that? When I ask myself that, what stands out most compellingly for me is the things others might do to me—things I clearly have reason to want them not to do. Now what is clear in my own case, I must accept in the case of others, since I am trying to find a basis for the criticism of morality that is neutral among people.

A reason for wanting morality, then, lies in the benefits we each derive from the moral restraint of others.[19] Our business now is to refine this talk of 'benefit'. Note first, it will not do simply to favour benefits; sometimes the choice is between benefits for one person and benefits for another. We need a criterion, neutral among people, for making the hard choices: what constitutes a fair and acceptable standard for adjudicating conflicts of interest? Utilitarians propose the standard that equal interests shall count equally—equal differences in benefit shall count equally. This utilitarian standard of adjudication has been widely attacked as ignoring matters of distribution, moral entitlement, and moral desert. Utilitarians will want to respond as Sidgwick does, by attempting to show that these various non-utilitarian considerations provide no coherent and acceptable standards for decision except as they can be justified on grounds of utility. That, however, is not something I shall attempt here; my question is what 'utility' has to be if it is to serve the rationale I am developing.

What shall count, then, as a benefit, and what shall count as equal differences of benefit? The classical utilitarians, I believe, supposed this question to be straightforwardly decidable. There is, they thought, recognizably such a thing as a person's 'good'. We can make full sense, they thought, of my 'acting for the good of someone else' or 'prefer-

[19] For non-utilitarian treatments of morality that use this starting-point, see Baier (1957) and Gauthier (1967).

ring my good to his'. A person's good is whatever is valuable simply for the way it impinges on him or on his life. His utility, then, is the degree to which his good is realized, the worth to the person himself of leading the life he leads. It is, I shall say, the *intrinsic reward* a person gets from the life he leads. The intrinsic reward of a life, they thought, is introspectable; with sufficient care, I can recognize how intrinsically rewarding my course of life is. True, it takes sensitivity and a clear head for relevant distinctions to achieve this recognition. Thus I can be in doubt as to whether the intrinsic reward of my life consists solely in pleasure and the absence of displeasure, and philosophers can dispute the issue. Such assessments, though, are intelligible in principle, however difficult they are to achieve.

VI. Interpersonal scrutability

How do questions of interpersonal comparability stand if there is such a thing as the introspectable intrinsic reward of living a life? Later I shall briefly take up the question we in the late twentieth century are likely to find more pressing: whether there is such a thing. In this section, though, I shall assume that there is, and discuss interpersonal comparability on that basis.

First we need to get straight the content of what I am assuming. It is not simply that we make introspective judgments of intrinsic reward. I find that evident enough: I sometimes judge that an experience is good or bad, simply because of what it is like to have such an experience, and I presume others do so too. The further assumption I am discussing is that these judgments are sometimes veridical: that they have a reasonable sense, that they can be true or false, and that if they are made carefully, sensitively, and with a clear head, they will normally be true and well-founded.

It is often said that scepticism of interpersonal comparisons of utility reduces to scepticism of the existence of other minds.[20] If my judgments of the intrinsic reward of my life are veridical, then something like that is indeed the case. For in that case, the problem of interpersonal comparisons is one of knowing what others experience, of knowing what it is like to be living their lives and experiencing things as they do. If I do not know what it is like to be living your life as you, then

[20] See Little (1957, Ch. 4).

I cannot compare the intrinsic reward of your life for you with the intrinsic reward of my life for me. For then although I can judge the intrinsic reward to me of the life I am leading, I cannot judge the intrinsic reward for you of the life you are leading. That is not because I cannot evaluate a life I understand, but because I do not know what the thing I am to evaluate is like. I do not know what your life is like for you.[21] If I knew that, I could make the interpersonal comparison of intrinsic reward.

Does it make sense, though, to talk of what it is like to be someone else, leading his life? I can imagine, more or less vividly, what it *might* be like to be someone else, leading his life and experiencing it as he experiences it. I can remember what life has been like for me at times when I was different in various respects from the way I am today. I can likewise imagine what it would be like to be different in various respects from the way I am now. To imagine what it would be like to experience life as someone else, I just have to extend my imagination to the point where I imagine myself to be different in more respects from the way I am now in fact, and to have a set of memories quite different from the ones I have in fact. The hard question is whether I can get matters right or wrong in such imaginings. How do I know if I have succeeded in imagining what it is like to be you and be leading your life? Here the soft-headed talk of imagining meets a hard-headed verificationist challenge. The question of whether I have got it all right, even in imagining you to have consciousness, is the problem of other minds. The question of whether I can correctly or incorrectly imagine those aspects of your life that make it more or less intrinsically rewarding to you to live it is the problem of interpersonal comparisons of utility. (All this, bear in mind, is still on the assumption that I can make veridical judgments of the intrinsic reward of a life as I imagine it.)

Here Harsanyi's (1955, Sect. 5) arguments can be brought to bear—though wrenched from their original context. Take first two hypothetical people who are exactly alike in every observable way. It would be gratuitous to suppose that they differ in unobservable ways, and so if

---

[21] My argument here essentially accepts the argument from analogy for the existence of other minds, and for more specific conclusions about their contents. Wittgenstein (1953) criticizes the arguments vigorously, and so I am in effect supposing that his arguments are wrong.

each imagines the other to experience life as he does, he imagines correctly: that is what canons of scientific method allow us to conclude. Take next two hypothetical people, one of whom was once exactly like the other in all observable respects, and who remembers what that was like. He is correct, we have already said, in supposing that the other experiences life as he used to, and so if his memory is to be trusted, and he imagines the other to experience life as he remembers experiencing it, then, we may trust, he imagines correctly.

The issue of whether memories of this kind can ever be trusted is, of course, a perplexing one, but this much can be said: such trust is needed for it to be at all possible for a person to learn from experience how to pursue his own good. Even a fool, a proverb implies, can learn from his own experience. Suppose he learns, though, from a painful fall from his bicycle, not to try to make sharp turns in wet weather at high speed. If his memory of what it was like to be hurt is not to be trusted, then, it would seem, the more fool he: if he has no reason to think it a bad thing to be hurt, then the danger of getting hurt gives him no reason to take turns slowly. If memories of experiences cannot be trusted at least to some extent, that vitiates not only interpersonal comparisons of utility, but much of what pertains to questions of how to lead life.

How do we get from fantastic cases of exactly similar people to comparisons among large varieties of actual people? In the first place, an individual learns not only what the intrinsic reward of particular segments of his life has been, but how the intrinsic reward of a part of his life depends on general features of that part of his life: on what he is like then, what he does, and what happens to him. What his experiences confirm or disconfirm, then, is hypotheses about how the intrinsic reward of a course of life depends on a range of combinations of features of the person and the life he leads. Call these *personal hypotheses*. It is indeed personal hypotheses that each of us must use for pursuing our own good. Since I can never expect to repeat a segment of my life exactly, I must predict the intrinsic reward of living in a particular way by the general features I expect it to have. The judgments of intrinsic reward here will be roughly cardinal: they will distinguish big differences in intrinsic reward from small, and do so finely enough to guide decisions of what is worth sacrificing now for what future gains and what is worth risking for what.

We must now ask whether these personal hypotheses can be combined, in principle, to form a well-confirmed interpersonal hypothesis. An *interpersonal hypothesis* will say how the intrinsic reward of anyone's life depends on what that person is like and the kind of life he leads. To say that such a hypothesis can be well confirmed in principle is not to say that we can hope, in the course of our lives, actually to confirm such a hypothesis. Even the most wise and sensitive person must probably go through life merely gathering a rich variety of hints of what life can be like for others who are quite different from him, and the greatest biography or novel will constitute an inadequate depiction of what life is like, or what life can be like, for a single person. The question here is not one of practicality, but of whether a strong, truly sceptical claim is correct: that not only will these hints always be compatible with a wide range of pictures of a person's inner life, but that these pictures could never converge, even with all the hints we could in principle obtain. True, the sceptic we are confronting may agree, when we know a person sufficiently well, we will have a sense of insight into what life is like for that person. That sense of insight, though, is merely a symptom of our lack of a capacity to visualize alternatives that are equally in accord with our evidence.

Now I have argued that to understand prudence, in the philosopher's sense of the rational pursuit of one's own good, we need cardinal comparisons of intrinsic reward within a single life. I have been assuming in this section, for the sake of discussion, that we can make such comparisons and communicate them. Strictly construed, then, what I am supposing is that we can communicate ratios of differences in intrinsic reward: the ratio of (α) the difference living life A as opposed to life A^* makes to the intrinsic reward of a person's life, to (β) the difference living life B as opposed to B^* makes to the intrinsic reward of his life. That is what a person needs in order to draw on his experience to plan his life with prudence. The claim of the sceptic who accepts what I have been assuming will be that these ratios are all that can be communicated.

What can be said to this sceptic? One answer to consider is this: A set of personal hypotheses determines a unique interpersonal hypothesis as long as there is enough interpersonal overlap. Personal hypotheses, we have been supposing, can be communicated up to a factor of scale. I can tell you what, on my personal hypothesis, I would find more rewarding intrinsically or less so, and compare magnitudes of differences. I cannot tell you on an absolute scale, though, how intrinsically rewarding my life is or how big a difference in intrinsic reward two different course of life would make to me. Speak now of each person's *personal hypothesis* in the singular: the most comprehensive of the personal hypotheses that are well confirmed for him in every respect. If the domains of your personal hypothesis and mine intersect at at least two points of different intrinsic reward, we can calibrate them. The domains intersect at a point if we can jointly imagine our being exactly alike, in the following sense. The point in question is a set of observable characteristics that a person and a part of his life might have: a certain kind of person leading a certain kind of life. The point is *within the domain* of my personal hypothesis if it is well confirmed how intrinsically rewarding I would find being that sort of person leading that sort of life. The domains of our personal hypothesis intersect at a point, of course, if the point is within the domain of each: if it is well confirmed, on our respective scales of intrinsic reward, how intrinsically rewarding each of us would find being that sort of person leading that kind of life. At the point of intersection, we are hypothetically identical, and so we may suppose that our lives have equal intrinsic reward; thus, at one level our personal scales of intrinsic reward are calibrated. Whether a set of personal hypotheses determines a unique interpersonal hypothesis, then, is a matter of whether there are at least two such points of calibration of different degrees of intrinsic reward.

This talk of calibrating fully confirmed personal hypotheses is at best, of course, an idealization, meant to illuminate what comparisons we might have genuine evidence for and against. The rationale for the idealization is roughly verificationist: if with full observational knowledge we could establish a comparison as fully credible, then limited observation can give us evidence for or against the comparison. If not, not. It is the limits of the power of possible evidence that are at stake here.

Return, then, to the idealization. A crucial step in the process remains to be elucidated. In the strict sense, the domains of the comprehensive personal hypotheses of two people never intersect. Any two people will differ in characteristics that cannot be altered. A man cannot experience what it is like to be a woman, and an adult whose early formative experiences were of one kind of upbringing cannot experience what it is like to be an adult whose early formative experiences were quite different. He can, perhaps, imagine what it *might* be like, but our question is still how to confirm that he is imagining what it *is* like. Here, following Harsanyi, we can say that it might turn out that changeable characteristics explain all we know about intrinsic reward, so that any observable personal differences that go with differences in unchangeable characteristics can be imputed to corresponding differences in changeable characteristics. Here we can apply Harsanyi's second canon of scientific method: 'when one variable is alleged to have a certain influence on another, the burden of proof lies on those who claim the existence of such an influence' (1955, Sect. 5).

Can all unchangeable characteristics be handled in this way? Surely not. Some unchangeable characteristics, it seems clear, affect experience in profound and complex ways. If, say, Christmas excites a person raised as a Christian and leaves cold a person raised as a Muslim that is probably because they had different formative experiences in early childhood. The childhood cannot be changed after the child's memories become clear enough for introspective comparisons of intrinsic reward to be possible.

Here, though, talk of comprehensive personal hypotheses hides the more piecemeal judgments we can confirm. I know in my own case that associations with early childhood have a special poignancy. The reports and expressive reactions of others may tell me that the same is true of them. I thus use both my own experience and observations of others to suggest and confirm modest interpersonal hypotheses about certain general features of the way the intrinsic reward of a life depends on its observable characteristics. In that way, I conclude that the Christian really is more excited by Christmas than is the Muslim.

There are, to be sure, alternative theories of psychic life consistent with observation. We might suppose the following: the Christian and the Muslim are equally moved by Christmas. The Muslim seems virtually unmoved, though, because he experiences life much more intensely than the Christian, but shows his feelings less. The feelings of each are intensified by early childhood associations, but that only brings the Christian's levels of feeling toward events with poignant childhood associations up to the level of the Muslim's normal feelings toward events. What is there to choose between this account and the claim that Christmas excites the Christian more than the Muslim?

The answer would seem to be that in the absence of any special evidence, it is gratuitous to attribute to the Christian a general intensity of experience less than that of the Muslim. On the original hypothesis, what is hypothesized is explained; on the alternative hypothesis, two further effects—greater general emotional responsiveness and less expressiveness on the part of the Muslim—are hypothesized gratuitously and left unexplained. That makes the original hypothesis antecedently far more plausible.

Where does that leave the problem of interpersonal comparisons of intrinsic reward? What it does is to open the possibility that such comparisons can be made. The sceptic thinks he can foreclose that possibility by showing that, given evidence consistent with one calibration of personal hypotheses, he can tell a story consistent with that evidence that involves a different calibration. What I have argued is that that move in itself does not settle matters; there may only be one story that does not introduce gratuitous elements—elements required by no evidence. In at least some cases, considerations of initial plausibility will pick out a single hypothesis of all those consistent with the evidence.

This case illustrates that any plausible interpersonal hypothesis must be embedded in a complex psychological picture of the nature and possibilities of human experience and the causes of differences in kinds of human experience. In confirming such a picture, we shall have to depend not only on conformity to the evidence, but on matters of simplicity and initial plausibility. In that respect, though, interpersonal comparisons of intrinsic reward will be far from unique in the annals of science. In a wide variety of scientific domains, a sceptic can appeal to a conventionalist strategy. For any hypothesis of how a theoretical magnitude depends on observable characteristics, the sceptic can offer dozens that have exactly the same observational consequences. Intrinsic reward—still on the assumption that it makes sense for alternative lives of a single person—is merely a case in point. The crucial question must be whether one of the interpersonal hypotheses consistent with observation uniquely well provides an initially plausible psychological explanation. The answer, then, must depend both on *a priori* questions of explanatoriness and initial plausibility and on *a posteriori* questions of which hypotheses are consistent with the evidence. In that respect, a science of intrinsic reward would be in no different situation from any other science groping to find its way.

It is important to remember that this section addresses a separable aspect of the problem of interpersonal comparisons of utility. That we may speak of the 'utility' of a single person in alternative courses of life, I have simply assumed—where 'utility' is now construed as 'intrinsic reward'. The problem in this section has not, at base, been whether the notion of 'intrinsic reward' is intelligible, but whether our lives are interpersonally scrutable. Can I meaningfully be said to be at least roughly correct, or to be wide of the mark, in imagining the experiences of someone else? That is the problem of scrutability, the interpersonal component of the problem of comparisons of intrinsic reward. So put, the question of scrutability is not specifically a question about 'intrinsic reward'. It is only together with another claim— namely, that there can be veridical introspective judgments of the intrinsic reward of a life—that complete scrutability would give complete interpersonal comparability of intrinsic reward.

VIII. Concluding remarks

The question of interpersonal comparisons of utility is most pressing as a question of the intelligibility of utilitarianism, and utilitarianism has a rationale, I have been claiming, only if we can make sense of the 'intrinsic reward' of a life, or something of the sort. That raises an intrapersonal question: whether we can, in principle, meaningfully compare the 'intrinsic reward' of alternative lives a given person might lead, or of different segments of the life he does lead. That question I have not addressed. What I have tried to do is to distinguish that question from the general question of interpersonal scrutability, and addressed the latter question on the assumption that intrapersonal comparisons of 'intrinsic reward' make sense. My conclusion is that no *a priori* argument will tell us the limits of interpersonal scrutability; the question is one for fundamental psychological investigation.

Difficult and profound as this question of interpersonal scrutability is, the more pressing questions, I suspect, are the ones I have not addressed. Do we have good reason to suppose that we can make such veridical introspective judgments of anything like 'intrinsic reward'? If not, what becomes of the utilitarian enterprise of Mill and Sidgwick? On these questions, I can here make only sketchy and, I fear, obscure remarks.

What happens if we deny the intelligibility of comparisons of intrinsic reward, even for a single person? Then unless we can find a satisfactory surrogate, interpersonal comparisons of intrinsic reward are also unintelligible. Indeed more than that is unintelligible. On their most evident construal, the concepts of rational egoism or rational self-interest become likewise unintelligible. For given an account of what constitutes the rational pursuit of a goal, we can define rational self-interest as the rational pursuit of intrinsic reward in one's life. On that definition, though, if intrinsic reward is unintelligible, then so is rational self-interest.

Why, then, doubt the intelligibility of 'intrinsic reward'? Precisely because there may be no ultimate sense in the question of what is best for me, when some hard choices must be made. Should I care intrinsically about my integrity, my independence of mind, or the truth of the beliefs that are central to my view of the universe? Or should I care only about my happiness? These questions, of course, need to be refined and pondered rather than merely broached, but the point here is that some of the reasons for denying the intelligibility of 'intrinsic reward' are precisely reasons for denying the intelligibility of rational self-interest: of the concept of clear grounds on which I could be advised how to act 'for my own good'. We could thus be sceptical of the meaningfulness of interpersonal comparisons of utility not because we think the experiences of others ultimately inscrutable, but because we are sceptical of there being any such thing as utility in general that could sum up the weight of good reasons even for a self-interested preference. This obstacle to interpersonal comparisons of utility, then, stems not from the comparisons' being interpersonal, but from their being comparisons of 'utility' in a sense that can provide grounds for preference. It was this part of the problem that Sidgwick took seriously.

The next move to explore might be the move back to preference: we may still want to define SI-preference in some way, and then identify the intrinsic reward of my life with the strength of my SI-preference for what happens over never having existed. I have borrowed from the work of others to suggest that that still does not yield an intelligible notion of self-interest, unless SI-preferences are unchangeable. If we could characterize them so that they both seemed to capture the notion of preferences based entirely on the intrinsic reward of one's own life and were unchangeable, that would amount to treating intrinsic reward as a secondary quality like colour: we pick it out by the subjective reaction it evokes. If, in addition, strengths of SI-preference were cardinally comparable for a single person, then the concept of intrinsic reward would be sufficiently well behaved for the discussion of the preceding section of this paper to apply: the remaining problems in principle would be ones of interpersonal scrutability. All that, however, would be a lot to expect.

If the problem with the intelligibility of utilitarianism is not so much that utilitarianism requires some interpersonal scrutability but that it requires a concept of a person's good or the intrinsic reward of his life, then much more is thrown into question than utilitarianism itself. For instance, a concept of valid moral rules as rules that would be chosen on grounds of rational prudence from behind a veil of ignorance would also be thrown into question.[22] The hope for a wide variety of normative theories must be that enough sense can be made of the concept of the intrinsic reward of a life to provide a satisfactory basis for ethics.

[22] Cf. Harsanyi (1955) and Rawls (1970). I owe much to Gauthier's attack on Harsanyi's argument; see Gauthier (1982).

REFERENCES

Baier, K. (1957) *The Moral Point of View*, Ithaca: Cornell University Press.

Bentham, J. (1789) *An Introduction to the Principles of Morals and Legislation* (1948), New York: Hafner.

Brandt, R. B. (1978) *A Theory of the Good and the Right*, Oxford: Oxford University Press.

Brandt, R. B. (1982) 'Two concepts of utility', in W. H. Williams and H. Miller (eds.), *The Limits of Utilitarianism*, Minneapolis: University of Minnesota Press.

Elster, J. (1979) *Ulysses and the Sirens*, Cambridge: Cambridge University Press.

Elster, J. (1983) *Sour Grapes*, Cambridge: Cambridge University Press.

Gauthier, D. (1967) 'Morality and advantage', *Philosophical Review* 76, 460–75.

Gauthier, D. (1982) 'On the refutation of utilitarianism', in W. H. Williams and H. Miller (eds.), *The Limits of Utilitarianism*, Minneapolis: University of Minnesota Press.

Harsanyi, J. C. (1955) 'Cardinal welfare, individualistic ethics, and interpersonal comparisons of utility', *Journal of Political Economy* 63, 309–21.

Hook, S. (ed.) (1967) *Human Values and Social Policy*, New York: New York University Press.

Little, I. M. D. (1957) *A Critique of Welfare Economics*, 2nd edn, Oxford: Oxford University Press.

Mill, J. S. (1859) *On Liberty*, London: Parker and Son.

Mill, J. S. (1863) *Utilitarianism*, London: Fontana Books.

Moore, G. E. (1912) *Ethics*, London: Williams and Norgate.

Overvold, M. (1980) 'Self-interest and the concept of self-sacrifice', *Canadian Journal of Philosophy* 10, 105–18.

Rawls, J. (1970) *A Theory of Justice*, Cambridge, Mass.: Harvard University Press.

Robbins, L. (1932) *An Essay on the Nature and Significance of Economic Science*, London: Macmillan.

Sidgwick, H. (1907) *The Methods of Ethics*, 7th edn, London: Macmillan.

Varian, H. (1978) *Microeconomic Analysis*, New York: E. W. Norton.

Wittgenstein, L. (1953) *Philosophical Investigations*, London: Macmillan.

7. Judging interpersonal interests

DONALD DAVIDSON

We constantly make judgments comparing the interests of two or more people. Sometimes such judgments provide reasons for action; sometimes they serve to explain or excuse the actions of the judge or of the agent whose interests are at stake. It is remarkable that we seldom find arriving at such judgments particularly difficult. Some cases are hard, of course; the same can be said concerning some decisions which affect one person only. But on the whole we do not experience the problem of comparing the interests of different people as being harder in kind or degree than comparing conflicting interests of our own. Naturally we are more apt to be ignorant of the interests of others than of our own; but this is a variable we have no trouble in accounting for.

Given how often we make judgments involving interpersonal comparisons of interest, and our general intuition that our moral principles and standards of rationality apply in much the same ways to such judgments as to personal decisions, it is strange that we find it difficult to explain how we make the interpersonal judgments. It is not a psychological explanation the absence of which is surprising; it is lacking, but not surprisingly. What is surprising is that we have no satisfactory view of the basis for our interpersonal comparisons. In this paper I argue that we have a basis for interpersonal comparisons, one that is actively and unavoidably at work. Its existence and nature are not necessarily obvious, though the facts to which I shall appeal are in some sense known to us all. I should perhaps say at the outset that it is not my aim to present a method or formula for deciding hard cases. Indeed, if my main thesis is correct, we should not expect to find a general method. At best what I am vaguely calling 'the basis' serves to give a point to discussions of alternative methods.

We are not clear what the basis for our interpersonal comparisons is;

neither is there a settled opinion as to the nature of such judgements. Early utilitarians believed that interpersonal comparisons were factual in character, and this view is again finding supporters. Others hold that judgments of interpersonal value are essentially normative.[1] Obviously the question of the basis for interpersonal comparisons is not independent of the normative-descriptive issue.

Some of the difficulty is no doubt due to differences about what are to count as judgments of interpersonal comparisons, what form they are to take, and what the elements of comparison are to be. Up to a point these matters may be settled by fiat, a task to which I now turn. In the end, however, deciding what the question is and addressing the question will turn out to be related in ways that defy dividing the discussion neatly into two parts.

I plan to concentrate on judgments made by a third party ('the judge') comparing the interests of two or more others. It will not do, of course, to rule that the values of the judge are not to enter into his judgment, since that would be to decide prior to understanding the nature of such judgments that they are not to be normative—whatever that means exactly. But we can decide without prejudice to central matters that the judgments shall not concern the interests of the judge. Of course we often do compare our interests with those of others; but these cases may introduce our own interests in two ways that are easy to confuse, once as part of what is judged and once as a determinant of the judgment. To keep these elements separate as far as possible, the judge, A, is to compare the interests of B and C, where A, B, and C are three different people. I take it for granted that A's judgment may depend, perhaps legitimately, on his own values or interests.

What are these interests or values the judge is to compare? The judge may be concerned, not with what B and C are interested in, or what they value or prefer, but with their 'true' interests, what would in some way be best for them, or perhaps best for society, or from the point of view of justice. Such comparisons, however, conspicuously and directly involve what the judge values or deems good or just, and these are matters I want to separate from the judgment as far as possible. So I choose to consider only those judgments that compare the preferences, desires or evaluations of those concerned. Let us say

---

[1] For a useful commentary on contemporary opinions, see A. K. Sen (1979).

that both *B* and *C* want to buy the same house. Each is prepared to buy it if he can. *A* makes judgments of these sorts: '*B* wants the house more than *C* does', 'According to their own preferences, *C* will gain more by getting the house than *B* will'; 'Considering their own evaluations, *B* will end up better off than *C* if *B* gets the house.' These judgments are different in important ways, and what might serve as a basis for one might not serve as the basis for another. But they are similar in that they compare, in various ways, the preferences, desires, or evaluations of *B* and *C*. It is such judgments that I wish to discuss.

I am aware of the dangers of lumping desires, evaluations, and preferences together, or of assuming that because they all enter into decisions and the formation of intentions they constitute a homogeneous group of motivational forces, But given the very general nature of the problem I am concerned with, I hope the distinctions among the evaluative attitudes (another attempt at a generic term) will not matter. In any event, I shall take no account of the differences, and shall use 'desire', 'evaluative judgment', and 'preference' to refer to the same broad set of attitudes.

Now suppose that our judge owns the house coveted by *B* and *C* and that he has decided to sell to one of them. We can imagine that there are three distinct steps in his reasoning, in so far as it involves the desires of *B* and *C*. First, he determines what he can of their preferences. Second, he compares these preferences, his judgment or judgments being of the kind mentioned two paragraphs back. Finally, he weighs these and any other factors he considers relevant, and makes a decision. If these steps really could be kept distinct in principle, then it would be clear that it is the second step with which we are concerned. It will turn out, however, that the first two steps are interdependent in a surprising way. Once this interdependence is described, it will be possible to improve the distinction between the last two steps, that is, to separate interpersonal comparisons from the evaluations based in whole or in part on them.

First, however, I want to point to an obvious difficulty in making the distinction. The difficulty is not superficial, because it tends to undermine a number of current theories about interpersonal comparisons. Let us ask the judge on what grounds he judges that *B* wants the house more than *C* does. The judge answers that although both would-be buyers have similar incomes and financial responsibilities, *B* is willing to pay more than C. (Obviously more subtle and perhaps convincing answers are possible.) The evidence or grounds of the judgment are then plainly factual and descriptive. But how about the judgment itself? If the judge goes on to use the judgment of comparative strengths of desire as a reason in favour of a decision to sell to B rather than to C, it is difficult not to count the judgment of interpersonal interests as a normative judgment belonging to the final evaluative step of deciding what course of action is best or most desirable. What the judge has done is use the fact that B is willing to pay more as a reason for increasing the value he sets on selling to B; the reference to strengths of desire merely acts as a middle term in getting from valued fact to valued action. The valued fact is that B is willing, under the circumstances, to pay more. Given the role in decision that the interpersonal comparison played, we are bound to say that when the judge held that B's desire was stronger than C's he was already making a value judgment.

The point does not depend on a connection with an action of such importance. Perhaps the judge merely expresses his view of the competing desires of B and C by way of a recommendation, an endorsement, or praise for a past act. If the judge supports these evaluative judgments by reference to his comparison of interpersonal interests, that comparison is infused with the explicitly normative character of the final evaluation. It may be suggested that to escape this problem we should consider cases where the judge does nothing about his interpersonal comparison. I will return to this idea presently. For the moment, though, it ought to be observed that actions taken on the basis of a judgment are often the only clear evidence of the nature and sincerity of the judgment.

Some writers insist, or acknowledge, that interpersonal comparisons are normative. Sen quotes Robbins to this effect.[2] Schick suggests that we 'assimilate' one utility scale to another by giving the same weights to the items highest and lowest on the individual scales.[3] Since he assumes interval measurement of individual preferences, this entails not only interpersonal comparisons of differences, but also absolute comparisons. He sees this as a just or moral way of treating people alike. Jeffrey has a somewhat more elastic proposal, which he also sees as normative from the start, though he believes it is essentially

[2] *Ibid.*, p. 190.　　　[3] F. Schick (1971).

the method we actually use in judging what is just or best.[4] I shall return to Jeffrey's proposal later.

These suggestions are not without interest, but they do not touch my problem. What I call a 'basis' for interpersonal comparisons cannot be something that is freely chosen, or that may be accepted by one person or society, but not by another. Schick's and Jeffrey's proposals concern what is just or fair, or what it is to treat someone as a person. If the concept of interpersonal comparison enters independently, it can only be if there is some further, non-arbitrary way of characterizing it, and this way has not been mentioned. Otherwise, these methods concern judgments of interpersonal strengths of preference only in that they tell us how we ought to judge conflicting claims.

Harsanyi has shown that if individual preferences and the social welfare function satisfy the von Neumann and Morgenstern axioms, and if social preferences are related to individual preferences in certain intuitively natural ways, then the weights the social welfare function assigns to various social states will be the sums of the individual preferences when the individual utility functions have been adjusted for origin and unit.[5] Jeffrey takes this to be a normative justification of a certain way of making interpersonal comparisons: our normative judgments that certain solutions to problems involving the interests of two or more people are fair show that we have used this method of making interpersonal comparisons. This interpretation of Harsanyi's result seems to me unexceptionable, though subject to the doubts I have already expressed if it is taken to provide a non-arbitrary basis for interpersonal comparisons. Harsanyi, however, has quite a different view of interpersonal comparisons. He says, '. . . interpersonal comparisons of utility are not value judgments based on some ethical or political postulates but rather are factual propositions based on certain principles of inductive logic'.[6]

Here is Harsanyi's argument, much simplified, but not, I think, distorted. It is in principle perfectly possible to make sound, justified, attributions of attitudes and beliefs to others. We do this on the basis of what can be observed of their behaviour, appearance or involuntary movements, what they say, and so forth. Unless we are to be sceptics about the minds of others, we have to allow that what can be observed

[4] R. C. Jeffrey (1971, 1974).

[5] J. C. Harsanyi (1955).   [6] *Ibid.*, p. 282.

is often adequate for knowledge of the interests, desires, intentions, worries and beliefs of others. But then it does not make sense to say that two people are alike in all relevant observable respects but have different thoughts and feelings. Or perhaps it makes sense, but it is bad science. To quote Harsanyi:

> If two objects or human beings show similar behaviour in *all* their relevant aspects open to observation, the assumption of some unobservable hidden difference between them must be regarded as a completely gratuitous hypothesis and one contrary to sound scientific method. . . . Thus in the case of persons with similar preferences and expressive reactions we are fully entitled to assume that they derive the same utilities from similar situations.[7]

In the last step, we determine the variables on which changes in mental states and attitudes depend, and so can compare the utilities of people who differ in taste, sensitivity and training. There is in effect one grand, empirical law of utility which relates the cardinal utilities of any individual to all the relevant variables. Though utilities will, as ever, be unique only up to a linear transformation, this will not matter to interpersonal comparisons, since everyone is on the same (interval) scale. There is a very similar argument in Waldner.[8]

It should be allowed that this argument has appeal, partly because of the strong analogies with legitimate arguments. There is, for example, G. E. Moore's observation that it would be absurd to say of two apples that one was good and the other not, but that there was no other difference. And it is certainly true that we often do know what others think and intend and want, and this knowledge must be based on what can be observed. It would be a mistake to deride this claim on the grounds that it is a form of behaviourism. Behaviourism is objectionable only if it maintains that mental states are nothing but the phenomena we normally take to be evidence for them; or that mental concepts can be explicitly defined in terms of the behavioural concepts. No such doctrine is involved here. But there is, I think, something wrong in Harsanyi's argument to show that interpersonal comparisons are 'factual propositions'. (I am not at this point questioning the conclusion.)

[7] *Ibid.*, p. 279.   [8] Ilmar Waldner (1972).

Let us accept as clear and correct the principle that if all possible evidence for two unobservables is the same, we should identify the unobservables. Plenty of questions could be asked about this principle, but they are not directly relevant to Harsanyi's argument. The question I want to raise concerns the application of the principle. Clearly it applies only to unobservables for which there is evidence. But is there any evidence for interpersonal comparisons? Since that is the issue to be settled, the answer can hardly be assumed as a premise of the argument. The argument seemed plausible because interpersonal comparisons are made (we are told) on the basis of the same sort of evidence we use for attributions of mental states generally. Since attributions of belief and desire to individuals are justified by the evidence, so must the interpersonal comparisons be. The cases are not in fact parallel. We do have evidence for legitimate attribution of belief, say. We can see this in a standard way by noting that a belief helps explain what is evidence for it. We use choice behaviour under certain circumstances as evidence for a belief. If the belief were different, it would not explain the same choice behaviour.

But now consider one of Harsanyi's examples. He says we might explain why one person is willing to work for lower wages than another at a given job by his having a lower disutility for labour if his physique is more robust and there is no ascertainable difference in economic needs. Would this explanation suffer from an arbitrary linear trans-formation of one utility scale? Clearly it would not. There is, on the one hand, the available evidence for each worker concerning relative strengths of preferences for work at various wages. Enough such evidence would permit a prediction for each worker, and hence a comparison of the wages at which each worker would choose to work. We could *call* this an interpersonal comparison of utility, but that would add no new information. If we discover a correlation between physiques and individual utility functions, that of course adds informa-tion that preferences alone do not yield. But the extra information yields exactly the same prediction, and explains the same behaviour, if the utility functions are transformed in permissible ways. Why, Har-sanyi asks, should we choose different absolute utilities for two people when the circumstances are the same in all relevant respects? The question is misleading. There is no motive for making any choice at all—unless we are to make judgments of merit or fairness or to suggest social policy, and then, I have urged, a straightforward normative judgment is involved.

Waldner points out, in effect, that there are other ways of obtaining interval scales of preference than by the von Neumann and Morgenstern method. (It would be better called the Ramsey method, since Ramsey came first, and unlike von Neumann and Morgenstern he did not assume that there are, or that people act on, objective probabilities.) One suggestion, at least as old as Bentham,[9] is to use least noticeable differences as a measure of equal differences in preferential strength. It is an open question whether, if one found satisfactory ways of determining utility by both the Ramsey method and the least noticeable differences approach, one function for an individual would be a linear transformation of the other. Suppose, to dream a bit, that it turned out to be the case that for each individual the functions could be made to coincide.[10] Would this, as Waldner thinks, justify us in making interpersonal comparisons of differences in utility? We would have an empirical way of making certain social choices, provided we decided to use least noticeable differences in making such choices. And we would have some general facts to which we could appeal to justify the choice. These are normative matters. But what is clear is that nothing new or different would be explained by normalizing utility functions by reference to a common unit based on least noticeable differences. Waldner says, '. . . the principle of not postulating any differences unless there is some reason to do so is hardly a matter under dispute'.[11] Perhaps so; but here there is no scientific reason to postulate anything at all.

I have argued that normative theories do not provide a basis for interpersonal comparisons on the ground that such theories are really decisions to use certain facts in the evaluation of social arrangements or decisions. This is not a complaint against such proposals, but a way of establishing what I mean by a 'basis' for comparisons. A basis would be non-arbitrary, at least in that it would not be chosen with an eye to subsequent value judgments. I have rejected a range of descriptive interpretations of interpersonal comparisons because comparisons interpreted in these ways add nothing to the descriptive or explanatory

---

[9] See Appendix IV, 555, in Baumgardt (1952).

[10] For a report on an experiment that tested a closely related hypothesis, see Davidson and Marschak (1959).

[11] Ilmar Waldner (1972).

power of theories to which they are added. Should we then give up on the attempt to found interpersonal comparisons? I think not. I think interpersonal comparisons have a basis in the sense that in the process of attributing propositional attitudes like beliefs, desires, and preferences to others interpersonal comparisons are necessarily made. The values that get compared are those of the person who attributes preferences or desires to someone else, and those of the person to whom the attributions are made. I do not mean that in attributing a value to another the attributor consciously or unconsciously makes a comparison, but that in the process of attribution the attributor must use his own values in a way that provides a basis for comparison.

It will be helpful to consider first how beliefs are attributed. Normal attributions of single beliefs are made against a background of knowledge or assumptions concerning the general character of a person's other beliefs, desires, and intentions, and a way of interpreting his speech. Under these favourable, though not unusual, circumstances, we often learn what someone believes through his assertions. Indeed, if we know that an assertion is sincere, and what the words, as spoken on that occasion, mean, we almost always will be right in supposing that the speaker believes what he has asserted. (The only exceptions will be cases where the speaker is wrong about what he believes.) Suppose someone asserts, and we deem him sincere, that pollux is a dread disease and castor is its cure; we conclude that he believes this. We will also conclude, without necessarily giving it a separate thought, that he believes that pollux is a dread disease; that there is a dread disease; that there is a cure; that something is either a cure or a poison; and so on. These further conclusions or assumings take no further time; they accompany the first conclusion. Yet they are separable from it, since it is possible for a person to believe something and fail to believe a logical consequence. Evidence might come along that would impel us to attribute a contradictory belief to a person. The more striking the contradiction, the harder it would be to be certain we were right; a flat enough contradiction in the beliefs as we interpreted them is sure to cast doubt on the interpretation. So if we are certain someone made two sincere assertions, and those assertions seem to contradict one another, we must wonder whether we are right about what was asserted. The doubt might easily go back to the question what the words used by the speaker meant.

Blatant logical error will tend to dissolve under good interpretation, but so will other errors to lesser degrees. If someone has a belief, whether true or false, about disease, this can only be because he has some views of what disease *is* that are true; otherwise there would be no reason to hold that it was disease he was thinking about. These other views in turn depend for their identity on further beliefs; and again, many must be true if some are false. Most beliefs are partial and conditional in the sense that we are less than certain of their truth, and the strength of each belief is a function of the strengths of further beliefs. Such dependencies reflect the fact that we take some propositions to be evidence for the truth of others. The moral for interpreters is that an acceptable interpretation must by and large reproduce the interpreter's pattern of conditional probabilities.

Similar remarks apply to more particular beliefs. Thus it must be part of what makes a belief a belief that a dog is before me that it is the presence and absence of dogs that causes the belief to wax and wane. As an interpreter, I can do no better at the start than to suppose that the content of someone else's belief is the same as a belief of mine that has the same cause. This is, of course, the crude outline of a policy for interpretation, and will necessarily be modified in many ways. The modifications depend on a large variety of matters: how well placed observers are, the condition of their sense organs, and what other relevant beliefs there is reason to attribute to them. Someone who, unlike me, does not believe dogs bark, may well fail to be caused to believe a dog is before him when I would be, given a dark night. The easiest errors to allow for in others are those we realize we might have made if we had been in their shoes.

To know the beliefs or other propositional attitudes of a person is to that extent to understand the person, for the propositional attitudes are by nature explanatory. That someone has a certain set of beliefs helps explain why he has, or acquires, others. Beliefs, in conjunction with preferences, explain intentions and intentional actions, as well as further preferences. If we know someone believes he over-tipped at a restaurant, and holds that such actions are shameful, we can explain why he is ashamed of having over-tipped.

The kind of explanation and understanding that flows from knowledge of the propositional attitudes of others involves, of course, the semantic properties of the propositions involved, and of the sentences that express them. This is why an interpreter cannot assign a certain attitude to someone else while supposing that the object of that attitude plays a radically different role in the thoughts of the other to what it plays in his own thoughts. So it would be a mistake to suppose that the accommodating policies I have been urging on interpreters are nothing but a matter of helping them discover truths whose character owes nothing to the thoughts of the interpreter. It is a tautology that we use our own concepts in understanding anything. What is not quite as easy to appreciate is that when it is the propositional attitudes of others we are trying to understand, two systems of thought must be made to mesh. This is not to suggest that the systems might be all that different; on the contrary, we see that comparisons with respect to similarities or differences depend on a large degree of similarity.

The process of making the beliefs and other propositional attitudes of others intelligible to ourselves necessarily involves our fitting others into our own scheme to some degree. There is no good sense, though, in thinking that this makes our attributions of attitudes less than objective, or that we are distorting the thoughts of others in the process of identifying them. These remarks would be appropriate only if we had another concept of what beliefs and desires and so forth are 'really' like to contrast with the one we apply in the way I describe. Failing such an alternative, we ought to accept the fact that the strategies we follow in making others intelligible are nothing but the process of determining what is true. The best interpretation an interpreter can devise, though it reads his own beliefs and standards of rationality into the minds of others, is as objectively correct as it can be.

Appreciation of the source of this objectivity can perhaps be encouraged by dwelling briefly on an aspect of the relation between communication and belief. If one aims to accomplish some purpose through saying something that is understood by a hearer, one must speak on the assumption that the hearer knows, or can learn, enough to interpret the speech as the speaker intends. This is impossible, however, unless the interpreter knows, with respect to many sentences of the speaker, that he holds them true (whether or not he utters them). The reason for this, put all too simply, is that the meanings of words are determined by the sentences containing those words that are held true. But then a successful interpreter of the utterance of a speaker knows both what the speaker's sentences mean, and, with respect to many sentences, that the speaker holds them true. Thus he knows, for each of these sentences, that the speaker has a particular belief. This shows not only how many beliefs must be as public as what is meant, but also, since what is meant must be intended to be public, that a speaker must intend that many of his beliefs be publicly discernible. Not only must meanings be social, but so must beliefs.

A speaker's beliefs must (in important and relevant matters) be known to an interpreter; but they must also be shared by him. The sharing takes two forms. First, as I have argued, there is no alternative to treating someone whose beliefs and meanings and other attitudes we wish to understand as being logically coherent. There can be intelligible exceptions, of course, given a general background of coherence. But I can no more understand the claim that someone constantly believes the conjunction of two propositions but not the conjuncts than I can make sense of an interpretation of his words that translates a certain expression by my 'and', but fails to endow it with the logical properties I do. Interpretation takes on a normative tone. In understanding others, in attributing propositional attitudes to them, I have no choice but to consider what inconsistencies do least harm to intelligibility; inconsistency here being inconsistency as I see it. My own standards of rationality necessarily enter the process of interpretation.

So do my views of what is true, both in general matters (what is the nature of disease, what counts as evidence for the presence of disease), and in matters of what is most plainly evident on occasion (is this a cat?). For as we have seen, it is only by treating or interpreting another as often or in strategically important ways agreeing with me that I make sense of (correctly interpret) his propositional attitudes.

Should we say that in interpreting others we 'compare' their logic and beliefs with our own? This seems to argue for two separate stages: first we learn what they believe, and then compare. This separation cannot, I have argued, be maintained until a general basis for interpretation has been laid down. Before conscious comparison is possible, our own standards of consistency and of the general character of the world have entered essentially into the process of determining what others think. A meaningful comparison depends on first having placed both minds in nearly enough the same realm of reason and the same material realm. So let us say that the attribution of propositional attitudes, while it involves a collating of beliefs, does not amount to a comparative judgment.

What has been said of beliefs applies with minor modifications to preference or desire. We make others intelligible by interpreting their beliefs and other attitudes; interpreting means assigning propositions to the propositional attitudes. Since the propositions we have available for assignment are ones available to us, and these are identified by their role in our own economies, a correct interpretation of someone else must make the objects of his or her attitudes the objects of corresponding attitudes of our own. This 'must' is one degree. The matching must be good enough in important respects to give a point to the failures of fit, these being the interesting cases where we disagree in what we hold true or in what we cherish. And there is the question what it means to say two attitudes correspond.

The same propositions are the objects of belief and of desire; they are also expressed by sentences. On the one hand, this reinforces the claim that the interpretation of desires proceeds along the same general lines as the interpretation of the cognitive attitudes. On the other hand, it guarantees that there will be multiple, and often conflicting, considerations in assigning an interpretation. We might know, for example, that a particular sentence apparently stands in certain logical relations to others for a given speaker; that he would prefer it true rather than some others; that his faith in its truth is modified to various degrees by observed changes in the world and by changes in his faith in the truth of other sentences. All of these considerations bear on the interpretation of the sentence, for on deciding it expresses a certain proposition, one has also decided on something believed and something desired; and the interpretation of further beliefs, sentences and desires has been much restricted. Clearly, different considerations will often point in different directions. Remembering that the underlying policy of choice requires us to choose the interpretation to match the other's beliefs and desires to our own, the conflict in considerations means that we have come across a recalcitrant case. Making a fair fit elsewhere perhaps forces us here to interpret a sentence the other holds true and wants true by one we hold true while hating that we must.

In the case of belief, two quite different pressures operate to form the beliefs of those we interpret into comprehensible patterns: a pressure in the direction of consistency (as the interpreter rates it) and a pressure in the direction of truth (also, of course, as the interpreter sees it). Two similar pressures are easy to recognize when it comes to identifying preferences or desires. We find it hard to credit straightforward examples of intransitivity of preference; we tend to explain apparent cases by changes in preference over time, or to suspect that the propositions being compared are larger in number than we had supposed. In fact all the axioms of Ramsey's theory of preference in the face of uncertainty exert a prima facie claim on interpretation; consistency of preferences with one another and with beliefs in Ramsey's sense is a constitutive pressure on interpretation simply because we cannot, on reflection, rationalize deviations from it. Deviations, especially if obvious, require an explanation, though often enough explanations are available, or must be assumed to be available.

The other pressure is towards agreement in preference. Preference can accept larger deviations than belief, but again there are limits. As with belief, the limits do not pick out particular attitudes where deviations are necessarily unintelligible; the limits are vague, just as degree of intelligibility can vary, and what they concern is the amount and kind of deviation that is acceptable. Disagreement on some general principles of evaluation is much harder to accommodate than on others, and disagreement on general principles is usually harder to make intelligible than disagreement on the value of particular acts and individual events. Nevertheless, with desire as with belief, there is a presumption (often overridden by other considerations) that similar causes beget similar evaluations in interpreter and interpreted. This is not, I should emphasize, either an empirical claim or an assumption for the sake of good science. It is a necessary condition of correct interpretation.

There should be no temptation to confuse the methodological need to interpret others as sharing our beliefs and values in important respects with the moral, or moralistic, advice to view the opinions and attitudes of others with imaginative sympathy, or to hunt for generous interpretations of the motives of others. These are policies we can follow or not, and following them depends on our having already largely made sense of what others believe and want. Nor is the method I claim we must follow in correct interpretation a matter of asking ourselves whether we would prefer to be in someone else's shoes, along with his ideas and preferences; I hope not, since I find such questions unintelligible. One might, in a metaphorical mood, describe my method for understanding someone else as putting him in my shoes; but this would certainly be a misleading metaphor if taken seriously. While plenty of imagination is called for in a good interpretation, I am not asking anyone to imagine he is playing another's part. I simply call attention to the fact that the propositions I must use to interpret the attitudes of another are defined by the roles they play in my thought and feelings and behaviour; therefore in interpretation they must play appropriately similar roles. It is a consequence of this fact that correct interpretation makes interpreter and interpreted share many strategically important beliefs and values.

Once we have decided what someone else believes or wants, and what his utterances mean, we can explain much of his behaviour. Would our explanation be as good if we were to make an arbitrary linear transformation of his utility (preference) function? Of course it would: this we have already seen. It is not explanatory power that causes us to put the preferences of others into relation with our own, but the process of deciding what the others' preferences are in the first place.

The 'basis' of interpersonal comparisons is then provided for each of us by his own central values, both his norms of consistency and of what is valuable in itself. These norms we do not choose, at least in any ordinary sense; they are what direct and explain our choices. So no judgment is involved in having one basis or another, much less a normative judgment.

In deciding how to interpret another there are surely, however, many options, for various interpretations are possible, given the same basis. There will be permissible alternatives which trade, say, simplicity for accuracy in detail, or stress consistency at the expense of correctness. This is just the 'indeterminancy of translation' extended to the more general case. If more than one interpretation of a speaker's words is possible, it is even more obvious that a theory that interprets his beliefs and desires along with his words has more degrees of freedom.

Is the determination of the desires of others relativistic because each interpreter starts with his own basis? I do not think so. Suppose two interpreters differ in their attributions of values to a third person. Before they can differ, they must interpret the words and beliefs of each other. In doing this each of necessity employs his own basis. For each, this supplies a ground of agreed-on standards, values and beliefs to which appeal can be made concerning their difference. Serious relativism provides no such common ground on which further intelligible discussion can be based.

Early in this essay I criticized a standard picture of how decisions are made concerning the interests of two or more people. According to this picture, we first decide what the interests of each person are; then we compare those interests in strength; then we judge or decide what should be done. I argued at the start that given this picture it was often difficult to separate the last two steps, and as a consequence the character of the interpersonal comparison was hard to determine. I have urged in the sequel that there is something fundamentally wrong with the idea that interpersonal comparisons can be isolated from simple attributions of desires or interests, since comparisons are implicit in such attributions. But of course what is implicit can be made explicit. There is no reason why we cannot judge the relative strengths of our own interests and those of others, or compare the interests of two others. My point has been that we do not have to establish, argue for, or opt for, a basis for such judgments. We already have it.

If it is true that the basis of interpersonal comparison already exists when we attribute desires to others, then we can, after all, make a clear distinction between interpersonal comparisons and normative judgments based on them. For issues of fairness, justice, and social welfare play no favoured role in our attributions to others of desires and preferences.

REFERENCES

Baumgardt, D. (1952) *Bentham and the Ethics of Today*, Princeton: Princeton University Press.

Davidson, D. and Marschak, J. (1959) 'Experimental tests of a stochastic decision theory', in C. W. Churchman and P. Ratoosh (eds.), *Measurement: Definitions and Theories*, New York: Wiley, pp. 223–69.

Harsanyi, J. C. (1955) 'Cardinal welfare, individualistic ethics and interpersonal comparisons of utility', *Journal of Political Economy* 63, 309–21. Reprinted in E. S. Phelps (ed.) (1973) *Economic Justice*, Harmondsworth: Penguin, pp. 266–85.

Jeffrey, R. C. (1971) 'On interpersonal utility theory', *Journal of Philosophy* 68, 647–56.

Jeffrey, R. C. (1974) 'Remarks on interpersonal utility theory', in S. Stenlund (ed.), *Logical Theory and Semantic Analysis*, Dordrecht: D. Reidel Publishing Co., pp. 35–44.

Schick, F. (1971) 'Beyond utilitarianism', *Journal of Philosophy* 68, 657–66.

Sen, A. K. (1979) 'Interpersonal comparisons of welfare', in M. Boskin (ed.), *Economics and Human Welfare*, New York: Academic Press.

Waldner, I. (1972) 'The empirical meaningfulness of interpersonal utility comparisons', *Journal of Philosophy* 69, 87–103.

8. Foundations of social choice theory: an epilogue

AMARTYA SEN

I. Introduction

A number of interesting questions have been raised about the foundations of social choice theory in this collection of essays presented at the
Ustaoset conference. As someone who could not attend the conference, I feel very privileged to have this opportunity of considering
and responding to some of the interesting ideas to be found in these
essays. Some authors, particularly Brian Barry and Jon Elster, are
deeply critical of what they see to be the foundations of social choice
theory. Others are more tolerant, even defensive, though they ask for
certain reforms, e.g. Goodin, Hylland. Still others—Davidson and
Gibbard—deal with social choice theory only indirectly, devoting
direct attention to specific problems of interpersonal comparisons of
utilities or interests, which may be relevant to assessing the foundations of social choice theory. A different framework for social choice
theory based on notions of exploitation rather than utilities is explored
by Roemer. There is much to discuss in all this, and this is what this
paper is about.

Let us begin at the beginning: what *is* social choice theory? This is
not an easy question to answer, and certainly the answer can be given
at quite different levels of specification. There is one important
contrast that should be aired straightaway, viz. that between (1) social
choice theory as *a field of study*, and (2) social choice theory as *a
particular approach or a collection of approaches* typically used in that
field of study. Some of the contributions in this collection take the
latter—narrower—view of social choice theory. Barry and Elster see
their criticisms as devastating for social choice theory because they find
these criticisms to be devastating to the established tradition. There

are really two issues here. I shall presently argue that these criticisms arise partly from misinterpreting the traditions of the subject, partly from inconclusive reasoning. But even if the criticisms had been devastating for the established *tradition*, the *subject* of social choice theory would not, of course, have been debunked. But what really makes these critical arguments interesting is not the grand claim of debunking social choice theory, but the possible validity of the arguments—to be examined—as critiques of the dominant traditions of social choice theory.

In this essay I shall be mostly concerned with the main traditions of social choice theory as they exist now rather than with the subject-matter of social choice theory as such. But I should make a few remarks on the subject as well. Social choice problems arise in aggregating the interests, or preferences, or judgments, or views, of different persons (or groups) in a particular society, and the exercise of aggregation can arise in very many different contexts, e.g. choosing social policies, or making social judgments, or assessing the state of the society. There can obviously be many different ways of dealing with an aggregation problem even when that problem has been precisely defined. Social choice theory as a field of study cannot be sensibly identified with only one (or a few) of these approaches, ignoring others. As a field of study, social choice theory has to have the same catholicity as would be associated with other fields of study such as moral philosophy, or monetary economics, or rural sociology. The demise of an approach used in a field of study has to be distinguished from the demise of that field of study. Monetary economic theory would still be there even when monetarism is laid to rest.

Considered as a field of study, there are many ambiguities as to what is or is not a part of social choice theory. Boundary questions arise in delineating any field of study, but I don't believe that it is particularly worthwhile to expend a lot of effort on boundary disputes. There is some arbitrariness in any such boundary specification, and it is partly a matter of convenience, partly of history. The usefulness of considering a field of study is not eliminated by doubts about precisely where the boundaries lie. (Disputes regarding the status of the Falkland Islands do not make it useless to consider the British and the Argentinians as two particular nations.) The delineation of social choice theory as a field of study has been influenced by historical causes as well as logical reasons, and in particular the concern with voting processes on the one hand and welfare economic judgments on the other has tended to give the subject a particular coverage that relates closely to the pioneering contributions of Kenneth Arrow (1951).

It is, however, useful to distinguish between various types of aggregation problems that are acknowledged to be part of social choice theory as a field of study. There are different ways of categorizing social choice problems, and I have tried to present elsewhere (Sen 1977a) some principles of classification. I shall not go through the principles or the categories again, but there is one particular distinction that is of immediate interest in examining the collection of papers included in this volume. The distinction is that between making systematic 'social welfare judgments' (e.g. 'This is, I think, best for society taking everything into account') and having institutional systems of social choice ('This wins over all alternatives in terms of rank–order voting—the accepted decision procedure').[1]

The papers of Davidson, Gibbard, Goodin, and Roemer are concerned mainly with social welfare judgments—or with the *ingredients* of such judgments—whereas the papers by Barry and Elster seem to be more concerned with choice procedures. Hylland deals explicitly with both and in particular examines the problem of judging social choice procedures by assessing their outcomes in the light of social welfare judgments.

In considering the arguments presented in the essays included in this volume, it is useful to have a brief look first at the main traditions of social choice theory as they have evolved. This is done—much too briefly for satisfaction—in the next section.

II. Arrow's format and variations

Arrow (1951) characterized the main problem of social choice as one of arriving at a social preference ordering R from an n-tuple of individual preference orderings $(R_1, \ldots R_n)$, one ordering for each person in a community of n people. The function f doing this transformation, $R = f(R_1, \ldots R_n)$, is called the social welfare function, and Arrow demanded that f should work for any n-tuple of individual preference orderings (unrestricted domain), satisfy the weak Pareto

[1] See Sen (1970a, 1977a) and Plott (1976).

principle, and be independent of irrelevant alternatives (so that the social ranking of any pair (x, y) must depend only on the n-tuple of individual rankings of *that* pair). He showed that with three or more alternative states and two or more (but not infinitely many) individuals, the only permissible social welfare functions must be dictatorial. While this astonishing 'General Possibility Theorem' (often called 'the impossibility theorem') caused most of the stir, important *positive* achievements included the specification of a particular format for social choice theory and the opening up of an entire area of study to systematic and formal treatment.

Arrow himself did not much explore the different *interpretations* of this format and in fact made many of the possible interpretations coincide with each other, through making specific assumptions, which happened to be quite unnecessary for his results but which he thought made sense in any case. Social preference can be interpreted in at least three different ways, viz. 'outcome evaluation', 'revealed preference relation of choice', and 'base relation of choice'. To explain, consider the strict social preference relation P. The statements xPy can be interpreted in at least one of the following three ways:[2]

(1) *Outcome evaluation*: x is judged to be a better state of affairs for society than y.

(2) *Revealed preference relation of choice*: Decision-making in the society should be so organized that y must not be chosen when x is available.[3]

(3) *Base relation of choice*: Decision-making in the society should be so organized that x must be chosen and y rejected in the choice exactly over the pair (x, y).

The interpretation of the social choice format developed by Arrow will vary with the chosen characterization of social preference, and we have the option of seeing the entire modelling in terms of social welfare judgments, or in terms of decision procedures of different types. The 'impossibility theorem' in particular will hold under each of these

---

[2] See Sen (1970a, 1977a, 1983a). Other interpretations are also explored there, including the distinction between 'descriptive' and 'normative' choice statements. Note also that the conventional use of the word 'preference' in this context must not be taken to mean that any *feeling of preferring* is necessarily involved. See also Little (1950) and Graaff (1957).

[3] In this formulation, it is implicitly assumed—what is usually explicitly stated (e.g. in Arrow 1963, pp. 13–17)—that from every non-empty set something or other *is* chosen, so that from the pair (x, y), x must be chosen, since y must not be.

interpretations if the Arrow conditions are defined correspondingly.[4] The same multiplicity of interpretations apply to the great variety of impossibility theorems and characterization results presented in the post-Arrow literature, involving weaker regularity properties.[5]

One particular common misunderstanding should be got out of the way before we proceed further. If we opt for a choice functional interpretation of the exercise and still use the binary relation of so-called 'social preference' R, it does not follow that we must of necessity assume that the choice function must be of the 'binary' type. Any choice function generates a revealed preference relation and a base relation, and we can talk directly in terms of these relations without having to make the further supposition that the choice function is itself binary (in the sense that the choice function can be recovered from the revealed preference relation or the base relation it generates[6]).[7] So the frequent assertions that Arrow's results and similar ones apply only to binary social choice procedures are simply not correct.

There are similar possibilities of alternative interpretations of individual preferences. We can also distinguish between the problem of aggregating individual *interests* (e.g. the conflicting claims of different persons in an income distributional problem with each pursuing his own interest) and that of aggregating individual judgments on some social matter (e.g. the different views of different people as to whether Britain should withdraw from the EEC).[8] Similarly, the individual preferences might be expressed by the persons themselves (e.g. by voting),[9] or guessed by someone doing the aggregation exercise (e.g. a

[4] See Sen (1970a, 1977a, 1982a, 1983a), Fishburn (1973, 1974), Blair, Bordes, Kelly, and Suzumura (1976), Ferejohn and Grether (1977), Matsumoto (1982). On related matters, see Hansson (1969), Schwartz (1970), Mas-Colell and Sonnenschein (1972), Brown (1973, 1975), Plott (1973), Binmore (1975), Blau and Deb (1976), Bordes (1976), Campbell (1976), Deb (1977), Monjardet (1979), Blair and Pollak (1982), Kelsey (1982).

[5] The literature on this is vast. I have tried to present and critically evaluate the main results in my draft chapter on 'social choice theory' (Sen 1979a) for *The Handbook of Mathematical Economics*, edited by K. J. Arrow and M. Intriligator, to be published.

[6] See Sen (1971), Fishburn (1974), Herzberger (1973), Schwartz (1976).

[7] This is not, however, to claim that the regularity properties of choice functions are invariably best stated in terms of the generated binary relations. The nature of the choice function may well be specified in terms of conditions that do not invoke either the base relation or the revealed preference relation, or indeed any generated binary relation.

[8] See Sen (1977a).

[9] The problems of preference relation and of correct social choice based on individual strategies have been studied for this case in a number of contributions; see Gibbard (1973, 1978), Pattanaik (1973, 1978), Satterthwaite (1975), Sengupta (1978), Laffont

Planning Commission arriving at a plan for the country by taking note of the interests of each group, or a person making a social welfare judgment by assessing what he sees to be the interests of different people).

Some of the perceived limitations of the format developed by Arrow for social choice theory are not, in fact, real, as should be clear from the above discussion. However, some other restrictions remain. First, the format is unable to take direct note of interpersonal comparisons of personal utilities or individual interests. This requires a different formulation of the problem, in terms of social welfare *functionals* SWFL (see Sen 1970a), but this is an easy extension of the Arrow format. A number of contributions have been made in recent years using interpersonal comparisons,[10] and this line of enquiry is now sufficiently well explored for it to count as a part of the mainstream of established social choice theory. The real difficulties lie not in formalizing the use of interpersonal comparisons but in interpreting their contents, procedures and legitimacy. Davidson and Gibbard have addressed this issue.

A second limitation of the original format is the tendency for it to rule out the use of non-utility information. This is not done directly, but the combination of unrestricted domain, independence,[11] and some condition like the Pareto principle has that effect (see Blau 1957; Arrow 1963; d'Aspremont and Gevers 1977). This implies that any two states that generate the same individual utilities must be treated in exactly the same way, no matter how much they may differ in their non-utility characteristics. This makes it difficult to take note of such notions as rights and liberty and also of such social characteristics as

(1979), Dasgupta, Hammond, and Maskin (1979), Barbera (1980), Dutta (1980), Moulin (1983), Suzumura (1983), Peleg (1984), and the literature referred to in these contributions.

[10] See, for example, Sen (1970a, 1970b, 1977b), Hammond (1976, 1979), Strasnick (1976), d'Aspremont and Gevers (1977), Deschamps and Gevers (1978), Maskin (1978), Gevers (1979), Roberts (1980a, 1980b), Blackorby, Donaldson, and Weymark (1984).

[11] The independence condition on its own has also been subjected to criticism for a long time. Recently there have been a number of interesting contributions using inter-profile conditions different from independence; see esp. Chichilnisky (1980). See also McManus (1975). On dispensing with inter-profile conditions altogether, see Parks (1976), Kemp and Ng (1977), Pollack (1979), Roberts (1980c), Rubinstein (1981). On 'positional' rules, see particularly Gärdenfors (1973), Fine and Fine (1974), and Young (1974).

exploitation or hunger. But once again, the welfarist character of the format can be removed easily enough by dropping one of the conditions that together yield the result, and by substituting instead some condition that brings non-utility information into play. One example is a condition of liberty, and the impossibility of combining even a minimal condition of liberty with the Pareto principle and unrestricted domain (as alleged in Sen 1970a) has been much discussed in this context.[12] Barry and Hylland have analysed this problem.

A more complicated issue in the use of non-utility information relates to incorporating *processes* of decision in the judging of outcomes or of actions, e.g. who exactly decides what the outcome should be in a particular case. This consideration is particularly relevant to the procedural formulation of liberty and rights in terms of valuing 'control' rather than the outcome, discussed by Nozick (1974), Sugden (1981) and others, and by Barry in this volume; see Section IV below.

Another complicated issue concerns 'counterfactual' considerations involving production conditions, e.g. those implicit in notions of marginal product (see Friedman 1953), or in the Marxian notions of exploitation (see Roemer 1982). The examination of the relevance of these issues for social choice theory is a fruitful new area of enquiry, to which Roemer has contributed in this volume.

This quick sequence of spotlights on related issues and problems has left out many important questions. I should emphasize that there has been no attempt here to 'survey' the literature. But the spotlighting may help in examining the nature and the value of the contributions presented at the Ustaoset conference. To that examination I now turn.

III. Well-being and interpersonal comparisons

It was remarked earlier that while Arrow's own format for social choice theory did not admit the use of interpersonal comparisons of

---

[12] Very many different types of results can be found in Sen (1970a, 1976, 1983a), Ng (1971), Batra and Pattanaik (1972), Bernholz (1974, 1980), Gibbard (1974), Nozick (1974), Blau (1975), Fine (1975), Seidl (1975), Campbell (1976), Farrell (1976), Aldrich (1977), Perelli-Minetti (1977), Breyer (1976), Ferejohn (1978), Kelly (1978), Stevens and Foster (1978), Suzumura (1978, 1983), Barnes (1980), Breyer and Gardner (1980), Breyer and Gigliotti (1980), Gardner (1980), Gaertner and Krüger (1981, 1983), Gärdenfors (1981), Hammond (1981), Sugden (1981), Austen-Smith (1982), Craven (1982), Levi (1982), Chapman (1983), Riley (1983), Wriglesworth (1983), among others.

well-being or interests, the format of social welfare functions can be easily extended to that of social welfare *functionals* admitting such comparisons (Sen 1970a). Many contributions have been made, in the last decade, in social choice theory involving systematic use of comparisons of welfare levels as well as welfare differences. I am, therefore, a little puzzled by Elster's remark:

> In the standard version, which is so far the only operational version of the theory, preferences are assumed to be purely ordinal, so that it is not possible for an individual to express the intensity of his preferences, nor for an outside observer to compare preference intensities across individuals (p. 105).

It is not obvious in what sense the literature on social welfare *functionals* is any less 'operational' than that of social welfare functions. It is certainly true that it is difficult to devise a procedure for taking note of interpersonal comparisons (or of cardinality) in *institutional* social choice procedures, but that is a characteristic of the world rather than of social choice theory as such.

The main use of social welfare *functionals* with interpersonal comparisons and cardinality[13] has been in the exercise of making social welfare judgments, e.g. by a planner, or a political observer, or indeed any person. This is a problem of exploring the appropriate frameworks for reflection, criticism and disputation.[14]

The real difficulties concern the actual possibility of interpersonal comparisons. Arrow (1951) expressed great scepticism about the meaningfulness and usability of such comparisons, and the difficulties of meaning and procedure are indeed important.[15] Davidson and Gibbard have made very substantial contributions to these questions in their essays in this volume. Davidson argues—convincingly—that

---

[13] It should be pointed out that the Arrow impossibility theorem can easily be extended to social welfare *functionals* admitting cardinal utility functions (taking note of 'intensities of preference') so long as interpersonal comparisons are not admitted; on this see Sen (1970a), Theorem 8*2. However, if the condition of independence is dropped, a class of interesting rules becomes admissible, e.g. the Nash social welfare function (see Sen 1970a, Theorem 8*1, and Kaneko and Nakaumura 1979).

[14] See Hylland's paper in this volume.

[15] See, however, Arrow's (1977) later use of interpersonal comparisons. See also Suppes (1966).

'the basis of interpersonal comparison already exists when we attribute desires to others', that 'comparisons are implicit in such attributions'. I shall not try to repeat the argument here, but if it is accepted, then the case for seeing social welfare judgments in terms of social welfare functionals with use of interpersonal comparisons is strong. And, as Davidson notes, it is also possible to distinguish between 'interpersonal comparisons and normative judgments based on them'. The last goes against the denunciation of attempts at descriptive interpersonal comparison (as opposed to making purely normative statements) presented by Robbins (1938) and others—a position that played rather an important part in basing modern welfare economics (and early social choice theory) on a format that did not admit any interpersonal comparisons as inputs in making social welfare judgments.

Gibbard argues for the necessity of a concept of 'intrinsic reward of life' in making normative judgments of the kind that utilitarianism makes. The real difficulty in the exercise lies, according to his analysis, not in making interpersonal comparisons of intrinsic rewards, but in having such a thing as a personal intrinsic reward. But if the concept of intrinsic reward is found to be unintelligible, then what becomes problematic is not just utilitarianism and other procedures of social welfare judgment based on interpersonal comparisons of well-being, but also such notions as rationality and prudence. There is scope for argument here, but I certainly would accept that in making interpersonal comparisons of intrinsic rewards of personal lives, the really difficult question is the *basis* of such attributions to persons rather than making interpersonal comparisons as such. In fact, Davidson's argument would have some application here. If the attribution of intrinsic worths of lives to others is intelligible, then some interpersonal comparison is already implicit in that exercise.

Davidson's and Gibbard's arguments have the effect of questioning the tradition of early social choice theory in basing social welfare judgments on non-comparable utilities or interests, admitting no interpersonal comparisons. I have discussed elsewhere the limitations and oddities of social choice frameworks without interpersonal comparisons, when dealing with social welfare judgments as opposed to institutional procedures of choice (see Sen 1970a, 1977b, 1979b), and the case for using informationally richer social welfare functionals was based on the advantages of the latter. But the problems of intelligibility and practicality of interpersonal comparisons are not easy to resolve.[16] Davidson's and Gibbard's contributions make the issues much clearer and in effect also make the case for using social welfare *functionals* stronger.

But the case for enriching social welfare frameworks with interpersonal comparisons must not be confused with the case for taking interpersonally comparable utilities as the only reasonable basis of social welfare judgments. Indeed, it can be argued that utilities provide a very limited basis of social welfare judgments even when they are interpersonally comparable. What is disputed in this assertion is 'welfarism', i.e. seeing social welfare as a function of the vector of individual utilities. It can be shown that even with interpersonal comparisons, welfarism fails to make adequate room for such principles as liberty, equality and justice.[17]

Two of the contributions in this volume take up these questions. Goodin's arguments for 'laundering preferences' relate to this issue of welfarism (though he presents his proposals in terms of refining utility information, rather than as incorporating non-utility information). He demonstrates the advantages of being discriminating about the role of different types of preferences and motivations. Goodin's arguments are quite persuasive, though they do rely a good deal on immediate intuitive appeal, as arguments in this type of exercise often do.[18] It is possible that Goodin has drawn too sharp a distinction between laundering preferences as opposed to using what he calls 'output filters'. If some preferences involving the violation of someone else's right (e.g. not to be tortured) is laundered out or discounted, then that process is not unlike blocking or countering the corresponding 'output' (the torture taking place). Laundering preference is a good way of seeing some of these issues, though there are other ways of characterizing the solutions proposed.

The laundering of preferences, though taking us beyond welfarism, cannot deal with the type of values that are reflected in Marxian considerations of exploitation. Roemer provides a far-reaching

---

[16] For different approaches used earlier, see Vickrey (1945), Little (1950), Harsanyi (1955), Suppes (1966), Kolm (1969), Gevers (1979), Sen (1979c), Blackorby, Donaldson, and Weymark (1984).

[17] See Sen (1970a, 1979b), Rawls (1971), Williams (1973), Phelps (1973), Nozick (1974), Dworkin (1977), Sen and Williams (1982), among others.

[18] See, however, Goodin (1982).

analysis of this question, building on his earlier work on different types of exploitation (Roemer 1982). This is not the occasion to try to assess the Marxian perspective *vis-à-vis* other approaches, but if the former approach to social welfare judgments is accepted, then the traditional models of social choice theory would require some substantial change. This is not because the formal structures of social welfare functions or functionals are incapable of incorporating the type of information on which Roemer's analysis focuses. Once welfarism (or 'strong neutrality', as it is sometimes called) is dropped, the descriptive information regarding states of affairs can be given a richer role, going well beyond counting only the utility information (see Sen 1977b). Indeed, as Roemer's analysis makes clear, some of the information in question is of the *counterfactual* type (e.g. what would happen if the workers withdraw from the arrangements, taking away their shares of means of production?). The condition of independence of irrelevant alternatives would have to be robustly violated to make room for using such counterfactual information regarding states of affairs that are not among the set from which choices are being made.

All this requires quite a departure from *early* social choice theory—indeed even from the dominant traditions of *recent* social choice theory. But formally the structure of social welfare functionals can admit procedures of the kind that Roemer outlines once the axioms of independence and Pareto principle are dropped. The real issues concern content and not just form, requiring the replacement of one set of axioms based on one set of information by another set of axioms using information of other types.

The foundations of social choice theory must permit these variations, since various different views of morality and social welfare judgment receive reasoned support. Social choice theory should be able to make room for the different approaches.

IV. Social choice and liberty

In his engaging paper in this volume, Brian Barry sets out to demonstrate that 'there is no incompatibility between liberalism and Pareto optimality' (p. 11), and in the process he identifies many 'mistakes' in social choice theory, including an 'error' that 'goes back to Kenneth Arrow's (1951) book', which seemed to have produced 'a logical monstrosity' (pp. 38–9). It is a difficult paper to discuss, since so many of the social choice concepts are so confidently misinterpreted in Barry's account of social choice theory that it is not easy to decide where to begin. It has been said that poetry communicates before it is understood. While reading Barry's paper, I felt that it is a serious failing of social choice theory that it does not share this amiable characteristic with poetry.

One difficulty arises from Barry's refusal to see that 'social preference' can be defined in choice functional terms, as was discussed in Section II above (and as indeed was explicitly spelled out by Arrow [1951]). Barry interprets 'socially preferred' or 'socially better' only in an oddly literal way and argues that 'liberalism . . . is not a doctrine about what constitutes a "socially better" state of affairs', but 'a doctrine about who has what rights to control what' (p. 15). I have argued elsewhere (Sen 1982b, 1983a) that this view of 'liberty as control' is fundamentally inadequate and this characterization of liberalism is correspondingly defective, but even if Barry's characterizations were fully accepted, the conflict between Paretianism and liberalism—thus defined—would remain completely unresolved. If a person does have a *right* to control a certain decision and he *wishes* to control it in a particular way, then requiring that outcome to emerge is a statement of 'social preference' under the choice–functional interpretation (see interpretation (2) in Sect. II). Indeed, this interpretation was explicitly discussed in the original presentation of the Pareto-liberal conflict (Sen 1970a, pp. 81–2), which Barry simply ignores.

In interpreting concepts such as 'social preferences', Barry chooses to confine himself to some very special interpretations, ignoring the definitions presented in the literature, and gets unaccountably excited about the folly of treating 'moral or aesthetic judgments' as 'preferences' (pp. 35–6). I have much sympathy for criticizing the terminology employed in social choice theory (see Sen 1979a, 1983a), but misinterpreting the language of social choice theory produces unreal battles that are purely verbal. The verbal welfare reaches a peak at the following remark of Barry:

> Why should I wish to insist not only that I should be able actually to decide what happens, e.g. whether my kitchen walls are pink or crimson, but also whether or not it will be 'socially better' that my

walls be pink rather than crimson? On the face of it, the latter has less to do with individual autonomy than with megalomania. Surely it ought to be enough that I can decide what colour my kitchen walls are (p. 34).

When 'I can decide what colour my kitchen walls are', there is in fact already a social preference relation (defined in terms of the choice function) congruent with what I personally choose. The substantial— as opposed to purely verbal—issues regarding liberty lie elsewhere.

Barry points to the possibility—treating it as rather a certainty—that Pareto-improving contracts will lead to Pareto optimality being attained along with each person's exercise of his rights. It is, of course, obvious that in any situation in which Pareto optimality is violated, there is a Pareto-improving contract that will restore Pareto optimality. The real question is whether it is reasonable to suppose that such Pareto-improving contracts must invariably take place and can be enforced. Oddly enough, Barry supposes that the only barrier to such contracts can arise from people *not being allowed* to enter into such contracts. He identifies what he sees as good grounds for disallowing such contracts in some cases (this is where *Doctor Fischer of Geneva or the Bomb Party* is invoked). But he does not believe that the case involving who should read *Lady Chatterley's Lover* provides good grounds for disallowing a contract by which Prude, who hates the book, has to read it in exchange for the book being *not* read by Lewd, who loves the book.

Barry may or may not be right in this particular judgment. I don't know, and don't much care, since the principal difficulties with the contractual solution lie elsewhere. The real issue is *not* whether Prude and Lewd will be 'allowed' to have such a contract,[19] but whether they will *seek* such a contract. I have discussed this issue elsewhere (Sen 1979d, 1983a), and here only quote a part of that discussion (Sen 1983a):

[19] Hylland in his paper distinguishes between 'alienable' and 'non-alienable' rights in this context. (See also Gibbard 1974, Farrell 1976, and Sen 1976, 1979d, 1983a.) Hylland makes a similar error to Barry's in supposing that the main issue is whether people are to be allowed to make such contracts. 'But the issue here is whether I, as an ethical observer, should criticize society for permitting them to strike a deal. I think not' (p. 67). This is a remoter question than whether the deal will be sought. Each party is also an 'ethical observer' himself, and has to face the question as to whether it would be right to offer the deal, or to accept it if offered.

It is important to note that the normative problems—both of *choice* and *outcome-evaluation*—may be viewed not merely from the position of outsiders, but also from the position of the involved individuals themselves. In that context, the individual's choice behaviour cannot—obviously—be taken as given. The question that has to be faced then is: 'Should I seek such a contract?' and not whether others have any reason to object if I were to seek such a contract. To try to 'solve' this problem by invoking one's preference as the great arbitrator is surely to beg an important moral question (p. 27).

Prude may refuse to offer the contract of reading a book that he hates to ensure that Lewd who loves the book must not read it. Lewd may refuse likewise. The fact that given their respective prudery and lewdness the contract looks attractive does not yet tell us how these individuals would act, since the normative question of how to act in these circumstances is not a trivial one. A person who is prudish in taste could be libertarian in values,[20] and refuse to get into a contract of this kind despite suffering from this refusal (through displeasure at Lewd's reading the book). The point is not that Prude and Lewd would never enter into such a contract. They might; but it cannot be presupposed that they must. Thus, this way of resolving the Pareto–liberty conflict is not a general solution.

There is a second difficulty. A contract of this kind is difficult—sometimes impossible—to enforce, and attempts at enforcement can be deeply alien to a liberal society. Since I have discussed this issue extensively elsewhere (1982b), I may confine myself here to quoting a part of that argument:

> what is certainly correct is that such a possibility of a Pareto-improving contract exists whenever the outcome is Pareto-inefficient; indeed that follows just from the definition of Pareto inefficiency. What is also true is that the outcome chosen individually on their own, e.g., Lewd's reading the book in the Lady Chatterley case, may be one of disequilibrium, if the individuals are in a position to have an agreement making Prude read the book and Lewd desist from it. (In fact, this possibility was explicitly pointed out in *Collective Choice and Social Welfare*, p. 84, at the time of

[20] Cf. Arrow (1951, pp. 17–19). See also Sen (1979c, pp. 479–87).

presenting the conflict.) But while this destabilizes the outcome with Lewd's reading the book, it does not make the result of the possible agreement—Prude's reading the book—a stable one either.

Suppose the agreement goes through. Prude now finds himself reading a book he hates, and there is every incentive for him to break the agreement on the sly. Also, Lewd finds himself kept away from reading a book that he would love to read—and he too has every incentive to break the contract. So the agreed Paretian contract is itself a disequilibrium outcome—just like the individually chosen outcome. There is, as it happens, no equilibrium outcome in this game. Every outcome is beaten by either individual action, or by joint collusive action. And that indeed *is* the paradox interpreted in terms of control.

So the contract-based outcome on which authors like Barry rely is not so much a 'solution' of the impossibility problem—it is in fact a *part* of the problem itself. The instability in the Lady Chatterley case is similar to the conflict faced in the so-called Prisoner's Dilemma. I shall not pursue here that game-theoretic analogy, but I should just note that in other examples of the impossibility of the Paretian liberal the structure of the game can be quite *different*. But in all these cases—whether or not they conform to the Prisoner's Dilemma—there is the common characteristic of every outcome being unstable, either due to individual action or due to joint action.

It is, of course, possible that such a cycle will be broken. One way of doing this is through compulsory enforcement of the contracted outcome. This raises deep moral questions as to whether it is right to have enforcement in an area of such private concern. Will it be right for a policeman—or some other 'enforcer'—to come and make sure that Prude is reading *Lady Chatterley's Lover* every morning? ('Doing my usual rounds, Sir', said the gentle policeman, 'and just dropped in to check that your eyes haven't deserted the good book, governor.') John Stuart Mill's strictures about people not having the freedom to sell their freedom is possibly relevant here.

But even aside from the moral permissibility of such contracts and—more important— of the binding enforcement of such contracts, there is the pragmatic question as to whether such contracts can, in fact, be practically enforced. Can it be made sure that Prude is in fact reading the book and not just *pretending* to? ('What', the gentle policeman again, 'was the last line you read, Sir? Do tell me, taking your time, of course.') There are far-reaching (and in my view, chilling) implications of trying to enforce contracts of this kind—involving the conduct of personal life. Once cannot help remarking that those who see in such contracts a method of ensuring the full exercise of individual liberty must have missed something about the nature of liberty (pp. 213–13).

The way of avoiding this enforcement problem is to develop behaviour patterns such that people voluntarily stick to the contracted actions (in this case Prude's reading the book he hates and Lewd's not reading the book he loves). The fact that these behaviour patterns would be contrary to the 'preferences' of the persons in question, is not in itself an insuperable difficulty, since counter-preferential behaviour is not necessarily alien to a social being.[21] But if counter-preferential behaviour is to be considered, then it is not at all clear why we should assume that Prude and Lewd must agree to offer or accept the contract in the first place. That assumption was questioned earlier in this essay (see also Sen 1979d), but Barry insists on completely preference-based choice in the behaviour of Prude and Lewd when it comes to making such a contract. Barry's belief in the case for having the contract seems to be based on his conviction that one would not 'choose to act like a damn fool and turn down a deal that would move me up my preference ordering' (p. 31). His position leads to the following dilemma: *Either* people must act according to their preferences, in which case the contract would be unstable and may break down; *or* they need not act according to their preferences, in which case we cannot be certain that the contract would be sought and made at all.

While the solution of the conflict between the Pareto principle and the condition of liberty through Pareto-improving contract does not stand, Barry's questioning of the nature of liberty as formalized in social choice theory is relevant and important. It is the same question that Robert Nozick (1974) had neatly posed in his earlier contribution:

---

[21] Counter-preferential behaviour will be impossible, indeed inconceivable, when preference is taken as the binary relation revealed by choice, but Barry's analysis is not based on that characterization of preference. Barry takes preference as a mental attitude guiding choice rather than being identified with choice. On various types of rationale for counter-preferential behaviour, see Sen (1973, 1977c), Baier (1977), Elster (1979, 1983), Majumdar (1980), Pattanaik (1980), van der Veen (1979), McPherson (1982), Akerlof (1983). Also Elster's paper in this volume.

A more appropriate view of individual rights is as follows. Individual rights are co-possible; each person may exercise his rights as he chooses. . . . Rights do not determine a social ordering but instead set the constraints within which a social choice is to be made, by excluding certain alternatives, fixing others, and so on (Nozick 1974, pp. 165–6).

Barry sees liberalism as 'a doctrine about who has what rights to control what' (p. 15). I would now like to argue against this conception of 'liberty as control'[22]—an argument I have presented elsewhere also. Once again I choose the lazy course of quoting—rather extensively —from that article (Sen 1982b):

I turn now to an examination of the control view of liberty. First a case that helps to bring out its rationale, especially in contrast with attempts to identify liberty with the pursuit of individual welfare, which Nozick—in my judgment rightly—criticizes. Lewd is confined to a hospital and asks you to go and get for him a copy of *Lady Chatterley's Lover*—unexpurgated edition—from the local library. You ask Lewd whether this is not the ideal time for him to read instead Shakespeare's plays, which Lewd has so far neglected to read. You opine that this would in fact be better for Lewd's own welfare, since Shakespeare will open doors for him. Lewd says that he absolutely agrees, and indeed reading Shakespeare would clearly be better for his personal welfare. Nevertheless, he would rather read Lawrence now, and would you please—as they say—do the needful. You do it, and get *LCL* for Lewd. There can be little doubt that this action serves Lewd's liberty, even if it does not serve his welfare. He chose *LCL*, and he exercised control over the decision as to what to read.

Consider now a variant of the case in which Lewd is in the hospital and you happen to know what his attitudes are about Lawrence and Shakespeare (including the fact that he would agree that Shakespeare would serve his welfare better), and that given the choice Lewd *would choose* to read *LCL* rather than *The Life of Timon of Athens*, and so forth. However, Lewd does not know that *LCL* is no longer on the banned list and even the local library has a copy for lending. Lewd simply asks you to get a book for him from the library. The control is yours. I would argue that even in this case Lewd's liberty is better served by your getting him *LCL*, which he *would have chosen if he had the choice*, rather than *The Life of Timon of Athens*. *LCL* is what Lewd would *prefer* to read and, given the chance, *choose* to read, and your giving him *Timon* instead would not serve what Isaiah Berlin calls 'a man's, or a people's, liberty to choose to live as they desire'.

Two disclaimers may be in order here. First, it is tempting to think that the case for your getting *LCL* for Lewd is really based on your regard for his *welfare* since *LCL* is what he would choose. . . . But this diagnosis is unsustainable since you happen to know that Lewd fully agrees that Shakespeare will indeed serve his welfare better, but despite that it is *LCL* that he would have chosen to read. The case for your getting *LCL* is based on what he *would have chosen* and not on what he and you think would serve his welfare better— quite the contrary.

Second, you might well decide that no matter what Lewd would have chosen, *Timon* is what he is going to get, since it is better for his welfare. That decision might well be morally defensible, but it would not be based on the consideration of Lewd's *liberty*. The point at issue concerns what course of action serves Lewd's liberty, and not how important Lewd's liberty should be in the overall decision regarding what you should do.

The concept of liberty involving such *counterfactual* reasoning (what the person *would have chosen*) I have called elsewhere *indirect* liberty. Indirect liberty lies outside the scope of liberty as control, but not outside the view of liberty as power. In being guided by Lewd's preference—what Lewd would have chosen—you are acknowledging that Lewd should have power over this decision, even if he does not happen to have the actual control over it. The characterization of implications of liberty in social choice theory does, of course, permit respecting indirect liberty—linking normative choice or judgment to the individual's preference and to what he or she would choose.

Society cannot typically be so organized that each person himself or herself commands all the levers of control over his or her personal life. The policeman helps my liberty in stopping mugging—presuming of course that I would choose not to be mugged if I had the choice—rather than giving me the control over *whether* to be mugged or not. Outcomes are also relevant for liberty, and not just the procedures of control.

Under the power view of liberty, we can rank the outcomes or states of affairs in terms of what the person involved would have chosen. Sometimes he would choose himself and thus some of these rankings will just reflect his own choice, and at other times he will not exercise the control himself but will still have power through the knowledge about his wishes. Thus, there is some real advantage in viewing liberty in terms of evaluation of overall states of affairs, and not just related to the single issue of procedure, to wit, who is actually exercising control. The characterization of liberty in social choice theory is, thus, not without merit.

Finally, incorporating liberty in the judgments of states of affairs also has the further advantage of being able to take a more informed view of liberty than the procedural control view—blind to the outcomes—can permit. Take the following example. Yesterday I spent the day at home. I could have of course done a number of other things, including calling on the village bore, conducting a census of the village cattle, and taking a dip in the village sewers. I chose not to do any of these other things, and I think that was an excellent choice. Suppose now I had been prevented by some village bully from exercising any choice. In Case I we assume that he asked me to stay at home and gave me no option. In Case II, he forced me to take a dip in the village sewers.

My liberty would have been affected in both cases—that is not in dispute. But is it correct to say that it would be *equally* violated in both cases? I think not. Although both would involve a violation of my liberty, the violation is clearly larger in Case II when I would be forced to take a dip in the village sewers than in Case I when I would be forced to stay at home, which—as it happens—I would have chosen to do anyway. While there is a loss of liberty in either case, it is absurd not to be able to discriminate between the two cases.

While there is some obvious advantage in seeing liberty as control, it is a mistake to see it *only* as control. The simpler social choice characterizations catch one aspect of liberty well (to wit: whether people are getting what they *would have chosen* if they had control), but miss another (to wit: who actually controlled the decision). But the view of liberty as control misses the former important aspect altogether even though it catches the latter. A more satisfactory theory of liberty in particular and rights in general would try to capture both aspects, and I have elsewhere (Sen 1982c, 1983b) tried to explore the possibilities of such a theory.

I end this over-long section with two general remarks. First, the importance of 'indirect liberty' (that of people getting what they *would have chosen* even though they are not able to control all the decisions themselves) relates to the nature of society. Given the complex interdependencies that operate in a society, tying together the lives of different people, it may be impossible to isolate their environment sufficiently to guarantee that each has all the controls over his or her personal life. The nature of physical and social environment makes parcelling out controls into self-contained bits deeply problematic, and the need to think in counterfactual terms as to what a person would have chosen *if he had control* becomes particularly important in that context.[23]

Second, the behavioural issues underlying the difficulties with the alleged contractual solution to the Pareto–liberty conflict raise questions that go well beyond considerations of liberty and the Pareto principle. For example, the possible refusal of Prude or Lewd to enter into a 'Pareto-improving contract' (making Prude read a book he hates and Lewd desist from reading one he would love to read) involve fundamental questions of values, including the value of the life one leads. This relates to Gibbard's notion of 'the intrinsic reward of a life', Roemer's concept of 'self-actualization',[24] and the idea of 'basic capabilities', which I have tried to pursue elsewhere (Sen 1980, 1982a, 1983c).

V. Political view and social choice theory

Jon Elster's wide-ranging paper is partly concerned with presenting a set of fundamental objections to social choice theory and partly with

[23] The importance of counterfactual considerations in other types of social reasoning has been penetratingly analysed by Elster (1978).
[24] The idea goes back at least to Marx, as Roemer notes, and has been explored by Maslow (1973).

exploring an alternative approach related to the works of Jürgen Habermas (and suggesting an approach 'in between these extremes'). Elster puts his basic objection to social choice theory thus:

> We can now state the objection to the political view underlying social choice theory. It is, basically, that it embodies a confusion between the kind of behaviour that is appropriate in the market place and that which is appropriate in the forum. The notion of consumer sovereignty is acceptable because, and to the extent that, the consumer chooses between courses of action that differ only in the way they affect him. In political choice situations, however, the citizen is asked to express his preferences over states that also differ in the way in which they affect other people. This means that there is no similar justification for the corresponding notion of the citizen's sovereignty, since other people may legitimately object to social choice governed by preferences that are defective in some of the ways I have mentioned. A social choice mechanism is capable of resolving the market failures that would result from unbridled consumer sovereignty, but as a way of redistributing welfare it is hopelessly inadequate. . . . the task of politics is not only to eliminate inefficiency, but also to create justice—a goal to which the aggregation of pre-political preferences is a quite incongruous means (p. 111).

I believe that this analysis is seriously flawed.

The preferences that serve as inputs in the social choice exercise need not be 'pre-political' ones. Indeed there is nothing to prevent social choice theory from taking on even those exercises that Elster sees as lying on the other side of the spectrum, to wit, the theory that Elster associates with Habermas's writings, for which 'the input of the social choice mechanism' is *not* 'the raw, quite possibly selfish or irrational preferences that operate in the market, but informed and other-regarding preferences' (p. 112). The format of social choice theory can be—and has been—used extensively to analyse aggregation problems with different types of preference inputs. There have also been several exercises contrasting the problems arising in the aggregation of self-centred preferences *vis-à-vis* those occurring in the aggregation of more other-regarding preferences (see, for example, Pattanaik 1971 and Sen 1977a).

In fact, a similar contrast to the one that Elster uses in *criticizing* social choice theory, had been used by Arrow himself, in *explaining* his theory, viz. that between 'the market mechanism' which 'takes into account only the ordering according to tastes' and using 'the ordering according to values which takes into account all the desires of the individual, including highly important socializing desires, and which is primarily relevant for the achievement of a social maximum' (Arrow 1951, p. 18). Arrow went out of his way 'to emphasize here that we must look at the entire system of values, including values about values, in seeking for a true general theory of social welfare' (p. 18).

Where social choice theory differs from a theory like the one Elster 'distills' from Habermas's writings is in its being more *parametric* in its approach. The individual preferences could be of various types—varying from being what Elster calls 'pre-political' to being intensely political, and from being totally self-centred to being immensely socially oriented. It is part of the generality of social choice theory to permit parametric variations of inputs as well as of outputs (see Sect. II above). For example, Arrow discussed, as one *special case*, exactly the 'market' type exercise that Elster seems to see as the only case that social choice theory can handle (Arrow 1951, Ch. VI: 'The Individualistic Assumptions'), but neither the other special cases nor the general structure (including such results as 'The General Possiblity Theorem') has this market-like character.

Elster explains that in terms of Habermas's theory, 'rather than aggregating or filtering preferences, the political system should be set up with a view to changing them by public debate and confrontation', and then 'there would not be any need for an aggregating mechanism, since a rational discussion would tend to produce unanimous preferences' (p. 112). I find this a rather Utopian view, as does Elster himself, judging from what he says elsewhere (pp. 115–16), and it is not really easy to see how antagonistic interests, including class conflicts, would all get submerged in 'unanimous preferences' merely by 'a rational discussion'. But no matter. If that is what would happen under the circumstances specified, then social choice theory, while still formally applicable, would take a rather trivial form.[25] The fact is,

---

[25] The case of unanimity corresponds to having a trivial 'domain' for the social welfare function. Domain restrictions have been much discussed in social choice theory in general terms; see, among others, Arrow (1951), Inada (1969), Sen (1970a), Pattanaik

however, that in the wretchedly divided world in which we live, social choice problems are not that trivial.

Part of Elster's problem may have arisen from a misinterpretation of the contents of what he calls 'given preferences' (pp. 105, 106, 111–12). While it is true that social welfare functions map 'given' individual preferences to social choice, this only pinpoints a *functional relation* that holds between individual preferences and social choice. It does not, of course, require that preference be 'given' *over time*. If education, or dialogue, or the political process, changes the individual preferences, then the inputs into the social choice mechanism must be seen as changing with the process in question. Not only is there nothing contradictory in this (the two senses of the word 'given' must not be confounded), but the format of social choice theory can even be used to derive a person's socially responsive preference as a function of pre-political preferences *and* the process of communication and social intercourse in the community.[26]

Elster's odd reading of 'given preferences' also allows him to see the following two sensible remarks, which he makes, as 'two sets of objections, both related to the assumption of given preferences', to social choices theory: 'I shall argue, first, that preferences people choose to express may not be a good guide to what they really prefer; and secondly that what they really prefer may in any case be a fragile foundation for social choice' (p. 106). In fact, formal social choice theory has done a great deal to expose the pitfall of ruling out strategic considerations and assuming that people's expressed preferences must coincide with what they really prefer (e.g. Dummett and Farquharson 1961; Gibbard 1973; Satterthwaite 1975; Pattanaik 1978; Schmeidler and Sonnenschien 1978; Laffont 1979; Moulin 1983; Peleg 1984), and 'the Gibbard-Satterthwaite theorem' is precisely addressed to Elster's first point. There is also an extensive literature on value-based counter-preferential choice (e.g. Sen 1977c; Majumdar 1980; Pattanaik 1980; Baigent 1980; McPherson 1982).

Regarding the second statement, it should be noted that the arguments for 'non-neutral' or 'non-welfarist' social choice procedures rest

---

(1971, 1978), Fishburn (1973), Kramer (1973), Salles (1975), Maskin (1976), Kalai and Muller (1977), Slutsky (1977), Grandmont (1978), Kelly (1978), Schofield (1978), McKelvey (1979), Kalai and Ritz (1980).

[26] See Sen (1970a, 1977b), Pattanaik (1971), Plott (1976).

precisely on the belief that social choice cannot be based exclusively on individual preferences ignoring other aspects of the states of affairs (see Sen 1970a, 1977b, 1979a, 1979b), and there is by now an extensive social choice theoretic literature on using non-utility information (see Sect. III and IV above).

Elster's criticism of social choice theory turns out to be inappropriate not because he is wrong in counting certain constraints as unjustified, but because he wrongly attributes these constraints to social choice theory. Hylland's paper provides a much better account of the content and traditions of social choice theory, and he explains some of the important roles that social choice theory can play (pp. 13–19).[27]

I still have a complaint or two about Hylland's otherwise excellent presentation. I have voiced one of them already regarding the interpretation of the Pareto–liberty conflict (fn. 19, Sect. IV above). Another issue relates a little to Elster's discussion, though Hylland's interpretation of individual preferences that serve as inputs to social choice mechanisms is very different from Elster's. Hylland writes about social choice theory.

> Nothing is said about motivation behind preferences. . . . Preferences can be based on egoistic concentration on private consumption, on altruism, on ideal principles of ethics, or on any combination of these factors. *We do not distinguish between these possibilities* (p. 53, italics added).

This is indeed so for many social choice exercises, especially those connected with characterizing *institutional* procedures for choice. On the other hand, in making social welfare judgments, 'the ethical observer', as Hylland calls a person making such a judgment, can certainly *distinguish* between different types of motivations underlying preferences. Motivational considerations have played an important part in, for example, the social choice literature on liberty, being either explicitly invoked (e.g. Sen 1970a, 1976; Fine 1975; Farrell 1976;

---

[27] For other examinations of possible roles of social choice theory, See Arrow (1951, 1963), Buchanan and Tullock (1962), Murakami (1968), Hansson (1969, 1976), Sen (1970a, 1977a), Fishburn (1971, 1973), Pattanaik (1971, 1978), Gottinger and Leinfellner (1978), Kelly (1978), Laffont (1979), Mueller (1979), Ng (1979), Feldman (1980), Pettit (1980), Sen and Williams (1982), Moulin (1983), Riley (1983), Suzumura (1983), Peleg (1984), Wriglesworth (1985), among other contributions.

Wriglesworth 1983), or indirectly used through associated characteristics (such as the 'dominance' property of self-regarding action identified by Gibbard 1974; Gaertner and Krüger 1981; Hammond 1981).

Hylland's generalization does, however, apply to a substantial part of social choice theory, especially the literature surrounding Arrow's General Possibility Theorem and also those dealing with such institutional procedures as the Borda method, majority rule, etc. The case for bringing in motivational considerations in the part of social choice theory that is concerned with ethical deliberation has always been strong, and recently they have been brought into formal models and informal discussions in a systematic way. Goodin's paper in this volume urges us to go further in that direction. In an indirect way, Elster too sends us that way. One of the effects of Roemer's arguments is also to push us in that direction, along with using other types of non-utility information.

VI. A concluding remark

In this 'epilogue' I have tried *both* to discuss the content and traditions of social choice theory *and* to respond to positive and negative arguments presented in the essays included in this volume. The variety of social choice problems and the need for discriminating treatment have to be borne in mind in assessing the entire discussion. For example, while the case for using social welfare functionals with interpersonal comparisons of utilities or interests is consolidated and strengthened by the contributions of Davidson and Gibbard (see Sect. III), the scope for this is obviously much more straightforward for ethical judgments than for institutional choices, both of which figure in social choice theory (see Sect. I). Similarly, the important lessons of Roemer's and Goodin's arguments about using non-utility information of particular types (see Sect. III and V) are immediately relevant to some types of social choice problems (e.g. in judgments of justice) in a way that they may not be to other types of problems.

The discussion of liberty and liberalism in Barry's paper provided the occasion both to defend the characterization of liberty in social choice theory and to present arguments for rejecting the view of 'liberty as control', independent of the outcomes (Sect. IV). It was also explained why the possibility of Pareto-improving contract does not resolve the conflict between liberty and Paretianism, no matter whether we accept the view of 'liberty as control' or choose an outcome-sensitive view of liberty. But once again the import of these considerations would vary depending on whether we are dealing with liberty-inclusive *ethical judgments*, or considering an *institutional structure* for ensuring liberty.

The fact that social choice theory deals with different classes of aggregation problems, with a variety of different types of inputs and outputs in the aggregation exercise, is an important thing to remember in examining the foundations of social choice theory. Elster's critique of social choice theory turns out to be flawed precisely because he tends to ignore this and relies on an account of social choice theory that simply does not fit. His otherwise illuminating discussion of the nature of the political problem gets wasted in a hopeless attempt at getting a telling critique of social choice theory out of it (Sect. V). Hylland, in contrast, takes note of the basic heterogeneity of social choice theory. His analysis also brings out how the same formal result (e.g. the so-called impossibility of the Paretian liberal) has very different types of relevance to different problems (e.g. ethical evaluation as opposed to devising an institutional structure for decentralized decision-making).

In examining the foundations of social choice theory, note must be taken *both* of the heterogeneity of the types of problems dealt with and the variations that the format permits in the nature of inputs, outputs and processes of aggregation. It is a mistake to think of social choice theory as a given set of complete ideas that are unleashed every time any problem is taken up for a 'social choice theoretic' treatment. 'Who does not distrust complete ideas?' W. B. Yeats has asked in his *Autobiography*. The answer is *not*, happily, 'the social choice theorist!'

REFERENCES

Akerlof, G. (1983) 'Loyalty filters', *American Economic Review* 73.
Aldrich, J. (1977) 'The dilemma of a Paretian liberal: some consequences of Sen's theorem', *Public Choice* 30.
Arrow, K. J. (1951) *Social Choice and Individual Values*, New York: Wiley.

Arrow, K. J. (1959) 'Rational choice functions and orderings', *Economica* 26.

Arrow, K. J. (1963) *Social Choice and Individual Values*, 2nd edn, New York: Wiley.

Arrow, K. J. (1977) 'Extended sympathy and the possibility of social choice', *American Economic Review* 67.

Austen-Smith, D. (1983) 'Restricted Pareto and rights', *Journal of Economic Theory*.

Baier, K. (1977) 'Rationality and morality', *Erkenntnis* 11.

Baigent, N. (1980) 'Social choice correspondences', *Recherches Economiques de Louvain* 46.

Barbera, S. (1980) 'A note on group strategy-proof decision schemes', *Econometrica* 47.

Barnes, J. (1980) 'Freedom, rationality and paradox', *Canadian Journal of Philosophy* 10.

Barry, B. (1965) *Political Argument*, London: Routledge.

Barry, B. (1970) *Sociologists, Economists and Democracy*, London: Macmillan.

Basu, K. (1979) *Revealed Preference of Governments*, Cambridge: Cambridge University Press.

Batra, R. N. and Pattanaik, P. K. (1972) 'On some suggestions for having non-binary social choice functions', *Theory and Decision* 3.

Bernholz, P. (1974) 'Is a Paretian liberal really impossible?' *Public Choice* 20.

Bernholz, P. (1980) 'A general social dilemma: profitable exchange and intransitive group preferences', *Zeitschrift für National-ökonimie* 40.

Binmore, K. (1975) 'An example in group preference', *Journal of Economic Theory* 10.

Blackorby, C., Donaldson, D., and Weymark, J. (1984) 'Social choice with interpersonal utility comparisons: A diagrammatic introduction', *International Economic Review* 25.

Blair, D. H., Bordes, G., Kelly, J. S. and Suzumura, K. (1976) 'Impossibility theorems without collective rationality', *Journal of Economic Theory* 13.

Blair, D. H. and Pollak, R. A. (1982) 'Acyclic collective choice rules', *Econometrica* 50.

Blau, J. H. (1957) 'The existence of a social welfare function', *Econometrica* 25.

Blau, J. H. (1975) 'Liberal values and independence', *Review of Economic Studies* 42.

Blau, J. H. and Deb, R. (1976) 'Social decision functions and veto', *Econometrica* 45.

Bordes, G. (1976) 'Consistency, rationality and collective choice', *Review of Economic Studies* 43.

Breyer, F. (1976) 'The liberal paradox, decisiveness over issues and domain restriction', *Zeitschrift für Nationalökonomie* 37.

Breyer, F. and Gardner, R. (1980) 'Liberal paradox, game equilibrium, and Gibbard optimum', *Public Choice* 35.

Breyer, F. and Gigliotti, G. A. (1980) 'Empathy and the respect for the right of others'. *Zeitschrift für Nationalökonomie* 40.

Brown, D. J. (1973) 'Acyclic choice', Cowles Foundation Discussion Paper, Yale University.

Brown, D. J. (1975) 'Aggregation of preferences', *Quarterly Journal of Economics* 89.

Buchanan, J. M. and Tullock, G. (1962) *The Calculus of Consent*, Ann Arbor: University of Michigan Press.

Campbell, D. E. (1976) 'Democratic preference functions', *Journal of Economic Theory* 12.

Campbell, D. E. (1978) 'Realization of choice functions', *Econometrica* 46.

Chapman, B. (1983) 'Rights as constraints: Nozick versus Sen', *Theory and Decision* 15.

Chichilnisky, G. (1980) 'Social aggregation and continuity', *Quarterly Journal of Economics* 94.

Craven, J. (1982) 'Liberalism and individual preferences', *Theory Decision* 14.

Dasgupta, P., Hammond, H. and Maskin, E. (1979) 'The implementation of social change rules: some general results on incentive compatibility', *Review of Economic Studies* 46.

d'Aspremont, C. and Gevers, L. (1977) 'Equity and informational basis of collective choice', *Review of Economic Studies* 46.

Deb, R. (1977) 'On Schwartz's rule', *Journal of Economic Theory* 16.

Deschamps, R. and Gevers, L. (1978) 'Leximin and utilitarian rules: a joint characterisation', *Journal of Economic Theory* 17.

Dummett, M. and Farquharson, R. (1961) 'Stability in voting', *Econometrica* 29.

Dutta, B. (1980) 'On the possibility of consistent voting procedures', *Review of Economic Studies* 47.

Dworkin, R. M. (1977) *Taking Rights Seriously*, London: Duckworth.

Elster, J. (1978) *Logic and Society*, Chichester: Wiley.

Elster, J. (1979) *Ulysses and the Sirens*, Cambridge: Cambridge University Press.

Elster, J. (1983) *Sour Grapes*, Cambridge: Cambridge University Press.

Farrell, M. J. (1976) 'Liberalism in the theory of social choice', *Review of Economic Studies* 43.

Feldman, A. M. (1980) *Welfare Economics and Social Choice Theory*, Boston: Martinus Nijhoff.

Ferejohn, J. A. (1978) 'The distribution of rights in society', in H. Gottinger and W. Leinfellner (eds.), *Decision Theory and Social Ethics: Issues in Social Choice*, Dordrecht: Reidel.

Ferejohn, J. A. and Grether, D. (1977) 'Weak path independence', *Journal of Economic Theory* 14.

Fine, B. (1975) 'Individual liberalism in a Paretian society', *Journal of Political Economy* 83.

Fine, B. and Fine, K. (1974) 'Social choice and individual ranking', *Review of Economic Studies* 41.

Fishburn, P. C. (1971) 'Should social choice be based on binary comparisons?', *Journal of Mathematical Sociology* 1.

Fishburn, P. C. (1973) *The Theory of Social Choice*, Princeton: Princeton University Press.

Fishburn, P. C. (1974) 'On collective rationality and a generalized impossibility theorem', *Review of Economic Studies* 41.

Friedman, M. (1953) *Essays in Positive Economics*, Chicago: University of Chicago Press.

Gaertner, W. and Krüger, L. (1981) 'Self-supporting preferences and individual rights: the possibility of Paretian libertarianism', *Economica* 47.

Gaertner, W. and Krüger, L. (1983) 'Alternative libertarian claims and Sen's paradox', *Theory and Decision* 15.

Gärdenfors, P. (1973) 'Positional voting functions', *Theory and Decisions* 4.

Gärdenfors, P. (1981) 'Rights, games and social choice', *Nous* 15.

Gardner, R. (1980) 'The strategic inconsistency of a Paretian liberal', *Public Choice* 35.

Gevers, L. (1979) 'On interpersonal comparability and social welfare orderings', *Econometrica* 47.

Gibbard, A. (1973) 'Manipulation of voting schemes: a general result', *Econometrica* 41.

Gibbard, A. (1974) 'A Pareto-consistent libertarian claim', *Journal of Economic Theory* 7.

Gibbard, A. (1978) 'Social decision, strategic behaviour and best outcomes', in H. Gottinger and W. Leinfellner (eds.), *Decision Theory and Social Ethics: Issues in Social Choice*, Dordrecht: Reidel.

Goodin, R. E. (1982) *Political Theory and Public Policy*, Chicago: University of Chicago Press.

Gottinger, H. and Leinfellner, W. (eds.) (1978) *Decision Theory and Social Ethics: Issues in Social Choice*, Dordrecht: Reidel.

Graaff, J. de V. (1957) *Theoretical Welfare Economics*, Cambridge: Cambridge University Press.

Grandmont, J. M. (1978) 'Intermediate preferences and the majority rule', *Econometrica* 46.

Hammond, P. J. (1976) 'Equity, Arrow's conditions and Rawls' difference principle', *Econometrica* 44.

Hammond, P. J. (1979) 'Equity in two-person situations: some consequences', *Econometrica* 47.

Hammond, P. J. (1981) 'Liberalism, independent rights and the Pareto principle', in L. J. Cohen, J. T'os, H. Pfeiffer and K. Podewski (eds.), *Logic, Methodology and Philosophy of Science*, Dordrecht: Reidel.

Hansson, B. (1969) 'Voting and group decision functions', *Synethese* 20.

Hansson, B. (1976) 'The existence of group preferences', *Public Choice* 28.

Harsanyi, J. (1955) 'Cardinal welfare, individualistic ethics and interpersonal comparisons of utility', *Journal of Political Economy* 63.

Herzberger, H. (1973) 'Ordinal preference and rational choice', *Econometrica* 41.

Inada, K. (1969) 'On the simple majority decision rule', *Econometrica* 37.

Kalai, E. and Muller, E. (1977) 'Characterizations of domains admitting non-dictatorial social welfare functions and non-manipulable voting procedures', *Journal of Economic Theory* 16.

Kalai, E. and Ritz, Z. (1980) 'Characterization of private alternative domains admitting Arrow social welfare functions', *Journal of Economic Theory* 22.

Kaneko, M. and Nakaumura, K. (1979) 'The Nash social welfare function', *Econometrica* 47.

Kelly, J. S. (1978) *Arrow Impossibility Theorems*, New York: Academic Press.

Kelsey, D. (1982) 'Acyclic social choice', D.Phil. thesis, Oxford University.

Kemp, M. C. and Ng, Y.-K. (1977) 'On the existence of social welfare functions, social orderings and social decision functions', *Economica* 44.

Kolm, S. Ch. (1969) 'The optimum production of social justice', in J. Margolis and H. Guitton (eds.), *Public Economics*, London: Macmillan.

Kramer, G. H. (1973) 'On a class of equilibrium conditions for majority rule', *Econometrica* 41.

Laffont, J.-J. (ed.), (1979) *Aggregation and Revelation of Preferences*, Amsterdam: North–Holland.

Levi, I. (1982) 'Liberty and welfare', in A. K. Sen and B. Williams (eds.), *Utilitarianism and Beyond*, Cambridge: Cambridge University Press.

Little, I. M. D. (1950) *A Critique of Welfare Economics*, Oxford: Clarendon Press.

McKelvey, R. D. (1979) 'General conditions for global intransitivities in a formal voting model', *Econometrica* 47.

McManus, M. (1975) 'Inter-tastes consistency in social welfare functions', in M. Parkin and A. R. Nobay (eds.). *Current Economic Problems*, Cambridge: Cambridge University Press.

McPherson, M. S. (1982) 'Mills moral theory and the problem of preference change', *Ethics* 92.

Majumdar, T. (1980) 'The rationality of changing choice', *Analyse und Kritik* 2.

Mas-Colell, A. and Sonnenschein, H. (1972) 'General possibility theorems for group decisions', *Review of Economic Studies* 39.

Maskin, E. (1976) 'Social welfare functions with restricted domain', mimeographed.

Maskin, E. (1978) 'A theorem on utilitarianism', *Review of Economic Studies* 45.

Maslow, A. H. (1973) 'Self-actualising people: a study of psychological health', in R. J. Lowry (ed.), *Dominance, Self-Esteem, Self-Actualization: Germinal Papers of A. H. Maslow*, Monterey, California: Brooks/Cole.

Matsumoto, Y. (1982) 'Choice functions: preference, consistency and neutrality', D.Phil. thesis, Oxford University.

Monjardet, B. (1979) 'Duality in the theory of social choice', in J.-J. Laffont (ed.), *Aggregation and Revelation of Preferences*, Amsterdam: North–Holland.

Moulin, H. (1983) *The Strategy of Social Choice*, Amsterdam: North Holland.

Mueller, D. C. (1979) *Public Choice*, Cambridge: Cambridge University Press.

Murakami, Y. (1968) *Logic and Social Choice*, New York: Dover.

Ng, Y.-K. (1971) 'The possibility of a Paretian liberal: impossibility theorems and cardinal utility', *Journal of Political Economy* 79.

Ng, Y.-K. (1979) *Welfare Economics*, London: Macmillan.

Ng. Y.-K. (1980) 'Welfarism: a defence against Sen's attack', *Economic Journal* 90.

Nozick, R. (1974) *Anarchy, State and Utopia*, Oxford: Blackwell.

Parks, R. P. (1976) 'An impossibility theorem for fixed preferences: a dictatorial Bergson–Samuelson social welfare function', *Review of Economic Studies* 43.

Pattanaik, P. K. (1971) *Voting and Collective Choice*, Cambridge: Cambridge University Press.

Pattanaik, P. K. (1973) 'On the stability of sincere voting situations', *Journal of Economic Theory* 6.

Pattanaik, P. K. (1978) *Strategy and Group Choice*, Amsterdam: North–Holland.

Pattanaik, P. K. (1980) 'A note on the rationality of becoming and revealed preference', *Analyse und Kritik* 2.

Peleg, B. (1978) 'Consistent voting systems', *Econometrica* 46.

Peleg, B. (1984) *Game Theoretic Analysis of Voting in Committees*, Cambridge: Cambridge University Press.

Perelli-Minetti, C. R. (1977) 'Nozick on Sen: a misunderstanding', *Theory and Decision* 8.

Pettit, P. (1980) *Judging Justice*, London: Routledge.

Phelps, E. S. (ed.) (1973) *Economic Justice*, Harmondsworth: Penguin.

Plott, C. R. (1973) 'Path independence, rationality and social choice', *Econometrica* 41.

Plott, C. R. (1976) 'Axiomatic social choice theory: an overview and interpretation', *American Journal of Political Science* 20.

Pollak, R. A. (1979) 'Bergson–Samuelson social welfare functions and the theory of social choice', *Quarterly Journal of Economics* 93.

Rawls, J. (1971) *A Theory of Justice*, Cambridge, Mass.: Harvard University Press.

Riley, J. M. (1983) 'Collective choice and individual liberty: A revisionist interpretation of J. S. Mill's utilitarianism', D. Phil. thesis, Oxford University; to be published by Cambridge University Press.

Robbins, L. (1938) 'Interpersonal comparisons of utility', *Economic Journal* 48.

Roberts, K. W. S. (1980a) 'Possibility theorems with interpersonally comparable welfare levels', *Review of Economic Studies* 47.

Roberts, K. W. S. (1980b) 'Interpersonal comparability and social choice theory', *Review of Economic Studies* 47.

Roberts, K. W. S. (1980c) 'Social choice theory: the single- and multiple-profile approaches', *Review of Economic Studies* 47.

Roemer, J. (1982) *A General Theory of Exploitation and Class*, Cambridge, Mass.: Harvard University Press.

Rubinstein, A. (1981) 'The single profile analogues to multiple profile theorems: mathematical logic's approach', mimeographed, Bell Laboratories.

Salles, M. (1975) 'A general possibility theorem for group decision rules with Pareto-transitivity', *Journal of Economic Theory* 11.

Samuelson, P. A. (1947) *Foundations of Economic Analysis*, Cambridge, Mass.: Harvard University Press.

Satterthwaite, M. A. (1975) 'Strategy-proofness and Arrow's conditions: existence and correspondence theorems for voting procedures and social welfare functions', *Journal of Economic Theory* 10.

Schmeidler, D. and Sonnenschein, H. (1978) 'Two proofs of the

Gibbard-Satterthwaite theorem on the possibility of a strategy-proof social choice function', in H. Gottinger and W. Leinfellner (eds.), *Decision Theory and Social Ethics: Issues in Social Choice*, Dordrecht: Reidel.

Schofield, N. (1978) 'Instability of simple dynamic games', *Review of Economic Studies* 40.

Schwartz, T. (1970) 'On the possibility of rational policy evaluation', *Theory and Decision* 1.

Schwartz, T. (1976) 'Choice functions, "Rationality" conditions, and variations of the weak axiom of revealed preference', *Journal of Economic Theory* 13.

Seidl, C. (1975) 'On liberal values', *Zeitschrift für Nationalökonomie* 35.

Sen, A. K. (1970a) *Collective Choice and Social Welfare*, San Francisco: Holden–Day; Amsterdam: North–Holland.

Sen, A. K. (1970b) 'Interpersonal aggregation and partial comparability', *Econometrica,* 38; reprinted in Sen (1982a).

Sen, A. K. (1971) 'Choice functions and revealed preference', *Review of Economic Studies*, 38; reprinted in Sen (1982a).

Sen, A. K. (1973) 'Behaviour and the concept of preference', *Economica* 40, 241–59.

Sen, A. K. (1976) 'Liberty, unanimity and rights', *Econometrica* 43; reprinted in Sen (1982a).

Sen, A. K. (1977a) 'Social choice theory: a re-examination', *Econometrica* 45; reprinted in Sen (1982a).

Sen, A. K. (1977b) 'On weights and measures: informational constraints in social welfare analysis', *Econometrica* 45; reprinted in Sen (1982a).

Sen, A. K. (1977c) 'Rational fools: a critique of the behavioural foundations of economic theory', *Philosophy and Public Affairs* 6, reprinted in Sen (1982a).

Sen, A. K. (1979a) 'Social choice theory', mimeographed draft chapter for K. J. Arrow and M. Intriligator (eds.), *Handbook of Mathematical Economics*, to be published by North–Holland, Amsterdam.

Sen, A. K. (1979b) 'Personal utilities and public judgment: or what's wrong with welfare economics?', *Economic Journal* 89; reprinted in Sen (1982a).

Sen, A. K. (1979c) 'Interpersonal comparisons of welfare', in M. Boskin (ed.), *Economics and Human Welfare*, New York: Academic Press; reprinted in Sen (1982a).

Sen, A. K. (1979d) 'Utilitarianism and welfarism', *Journal of Philosophy* 76.

Sen, A. K. (1980) 'Equality of what?' in S. McMurrin (ed.), *Tanner Lectures on Human Values*, Cambridge: Cambridge University Press; reprinted in Sen (1982a).

Sen, A. K. (1982a) *Choice, Welfare and Measurement*, Oxford: Blackwell.

Sen, A. K. (1982b) 'Liberty as control: an appraisal', *Midwest Studies in Philosophy* 8.

Sen, A. K. (1982c) 'Rights and agency', *Philosophy and Public Affairs* 11.

Sen, A. K. (1983a) 'Liberty and social choice', *Journal of Philosophy* 80.

Sen, A. K. (1983b) 'Evaluator relativity and consequential evaluation', *Philosophy and Public Affairs* 12.

Sen, A. K. (1983c) 'Poor, relatively speaking', *Oxford Economic Papers* 35.

Sen, A. and Williams, B. (eds.), (1982) *Utilitarianism and Beyond*, Cambridge: Cambridge University Press.

Sengupta, M. (1978) 'On a difficulty in the analysis of strategic voting', *Econometrica* 46.

Slutsky, S. (1977) 'A characterisation of societies with consistent decision', *Review of Economic Studies* 44.

Stevens, D. M. and Foster, J. E. (1978) 'The possibility of democratic pluralism', *Econometrica* 45.

Strasnick, S. (1976) 'Social choice theory and the derivation of Rawls' difference principle', *Journal of Philosophy* 73.

Sugden, R. (1981) *The Political Economy of Public Choice*, Oxford: Martin Robertson.

Suppes, P. (1966) 'Some formal models of grading principles', *Synthese* 6; reprinted in his *Studies in the Methodology and Foundations of Science*, Dordrecht: Reidel, 1969.

Suzumura, K. (1978) 'On the consistency of libertarian claims', *Review of Economic Studies* 45.

Suzumura, K. (1983) *Rational Choice, Collective Decisions and Social Welfare*, Cambridge: Cambridge University Press.

Van der Veen, R. J. (1979) 'Meta-ranking and collective optimality', *Social Science Information* 20.

Vickrey, W. (1945) 'Measuring marginal utility by reactions to risk', *Econometrica* 13.

Williams, B. (1973) 'A critique of utilitarianism', in J. Smart and B. Williams, *Utilitarianism: For and Against*, Cambridge: Cambridge University Press.

Wriglesworth, J. (1985) *Libertarian conflicts in social choice*, Cambridge: Cambridge University Press.

Young, H. P. (1974) 'An axiomatization of Borda's rule', *Journal of Economic Theory* 9.

Index